D0915396

TRIASSIC-JURASSIC RIFT-BASIN SEDIMENTOLOGY

TRIASSIC-JURASSIC RIFT-BASIN SEDIMENTOLOGY

History and Methods

John C. Lorenz

VNR Van Nostrand Reinhold Company
_____ New York

Library of Congress Catalog Card Number 87–6138

ISBN 0-442-26041-5

Printed in the United States of America.

Designed by Karin Batten

Van Nostrand Reinhold Company Inc.
115 Fifth Avenue
New York, New York 10003

Van Nostrand Reinhold Company Limited
Molly Millars Lane
Wokingham, Berkshire RG11 2PY, England

Van Nostrand Reinhold
480 La Trobe Street
Melbourne, Victoria 3000, Australia

Macmillan of Canada
Division of Canada Publishing Corporation
164 Commander Boulevard
Agincourt, Ontario M1S 3C7, Canada

16   15   14   13   12   11   10   9   8   7   6   5   4   3   2   1

Library of Congress Cataloging-in-Publication Data
Lorenz, J. (John)
  Triassic-Jurassic rift basin sedimentology.
  Bibliography: p.
  Includes index.
  1. Rocks, Sedimentary.   2. Rifts (Geology)   3. Geology, Stratigraphic—Triassic.
4. Geology, Stratigraphic—Jurassic.   I. Title.
QE471.L67   1987   551.7'62'097462   87-6138
ISBN 0-442-26041-5

With pleasure and heartfelt sentiment, I would like to dedicate this book to Lizzie. While I was out of circulation, she remained ever supportive—although I suspect that, at times, she would have preferred that I still swept streets for a living.

Contents

FOREWORD   xiii

PREFACE   xv

ACKNOWLEDGMENTS   xvii

**PART I: EVOLUTION OF CONCEPTS AND
INTERPRETATIONS OF THE HARTFORD AND
DEERFIELD BASINS**   1

INTRODUCTION

    Usefulness of the Historical Perspective   3
    Review of the Newark Supergroup in the Hartford
       and Deerfield Basins   6

1/Early 1800s   **15**

    Silliman and Maclure: Earliest Observations,
       European Roots   15
    Hitchcock: Ichnology and Early Sedimentology   19
    Lyell: The Bay of Fundy Analogue   29
    Rogers: The Theory of Primary Depositional Dip   29
    Percival: The Geology Map of Connecticut   33
    Redfield: The Age of the Deposits   35

2/Late 1800s   **37**

    Dana: Equivocal Interpretations and the *Manual of Geology*   37
    Russell: Broad Estuaries and Red Color   39
    Davis: Structure and Nonmarine Deposition   45
    Emerson: The Geological Map of Massachusetts and
       Marine Interpretations   49
    Hobbs: The Pomperaug Outlier   52

3/Early 1900s   **54**

Early Work   54
Barrell: Models of Nonmarine Deposition   55
Lull: Vertebrate Paleontology   59
Emerson: Holdover of the Estuarine Model   60
Longwell, Russell, and Foye: Sedimentology of Fan Deposits
    and Proof of the Eastern Border Fault   62
Bain and Wheeler: More Observations on Basin Structure,
    with Asides on Sedimentology   66
Redbed Controversy   67
Krynine: Petrography, Climate, and Redbeds   68

4/1950 to the Present   **75**

Mappers and Stratigraphers   76
Miscellaneous Studies of the 1960s: Structure and Basalt   80
Wessel, Sanders, and Klein: Applications of "Modern"
    Sedimentology   82
Advances in Other Areas   85
Palynology: Age and Depositional Environment   86
Fluvial and Lacustrine Sedimentation   88
Recent and Ongoing Work   95

SUMMARY   98

The Course of Progress   98
Gaps in Present Knowledge   100
Postscript   103

**PART II: SEDIMENTOLOGY OF TRIASSIC-JURASSIC
RIFT-BASIN DEPOSITS**   **105**

INTRODUCTION   107

Scope and Purpose   107
Regional Setting   111

5/Alluvial-Fan Deposits   **115**

  Alluvial-Fan Subenvironments   116
  Sieve and Debris-flow Deposits   118
  Midfan Fluvial and Sheetflood Deposits   120
  Fan-toe Sheetflood Deposits   125
  Nonstandard Alluvial Fans   130
  Facies: Patterns and Controls   132
  Summary   137
  Data Collection and Interpretation Techniques Illustrated by
     Studies of Rift-Basin Alluvial-Fan Deposits   137

6/Fluvial Deposits   **140**

  Braided-river Deposits   141
  Meandering-river Deposits   148
  Overbank Deposits and Postdepositional Modifications   150
  Facies: Patterns and Controls   157
  Summary   162
  Data Collection and Interpretation Techniques Illustrated by
     Studies of Rift-Basin Fluvial Deposits   162

7/Paludal Deposits   **165**

  Coal Deposits of the Deep River Basin   168
  Comparison with Other Triassic Rift-Basin Coals   173
  Facies: Patterns and Controls   176
  Summary   178
  Data Collection and Interpretation Techniques Illustrated by
     Studies of Rift-Basin Paludal Deposits   178

8/Lacustrine Deposits   **180**

  Shoreline Deposits   183
  Shallow-water Deposits   187
  Deep-water or Offshore Deposits   193
  Geochemistry of Lake Waters   198

Facies: Patterns and Controls   200
Evidence from Basalts   205
Summary   207
Data Collection and Interpretation Techniques Illustrated by
    Studies of Rift-Basin Lacustrine Deposits   207

9/Playa Deposits   **210**

Primary Playa Deposits   211
Secondary Modifications to Playa Deposits   213
Facies: Patterns and Controls   216
Summary   216
Data Collection and Interpretation Techniques Illustrated by
    Studies of Rift-Basin Playa Deposits   219

10/Marine-influenced Deposits   **221**

Moroccan Salt Basins   223
English Salt Basins   231
Aquitaine Salt Basin   242
Marine Influence in the Central East Greenland Basin   244
Facies: Patterns and Controls   244
Summary   245
Data Collection and Interpretation Techniques Illustrated by
    Studies of Rift-Basin Marine-influenced Deposits   245

11/Eolian Deposits   **248**

Eolian Deposits of the Fundy Basin   250
Eolian Deposits in England and Morocco   253
Interdune Deposits   255
Facies: Patterns and Controls   257
Summary   257
Data Collection and Interpretation Techniques Illustrated by
    Studies of Rift-Basin Eolian Deposits   258

12/Overview of Triassic-Jurassic Rift-Basin Sedimentology **260**

   Regional Controls on Rift-Basin Sedimentology 260
   Data Collection and Analysis 267
   Features of Modern Environments that Are Preserved in
      Rift-Basin Deposits 275
   Scientific Methods for a Historical Science 281

REFERENCES, PART I 287

REFERENCES, PART II 297

INDEX 309

Foreword

More than fifteen Triassic-Jurassic rift basins lie along a belt stretching from Nova Scotia to South Carolina, and nearly as many are in the subsurface to the east. Geological events of the past two decades have focused considerable attention on these Newark basins and their nonmarine deposits. Recent reconstructions of the Mesozoic opening of the Atlantic Ocean basin, for example, assigned the Newark troughs an important initial role in a sea-floor spreading scenario. Concurrently, concern about future petroleum reserves induced active exploration of these nonmarine basin deposits as potential sources of oil and gas.

In response to this awakened interest and expanded investigation, there have been numerous Newark and rift-basin symposia, as well as an outpouring of scientific papers on various aspects of the research. In addition, field trips to some of the basins have been laid out in guide books. These efforts have made the Newark basins and their deposits well-known features of North American geology; they are now significant subjects in any general geological text or curriculum. Most of the current studies have dwelt on new information and recent developments: little attention has been given to the long and revealing story recorded in earlier interpretations, or to the way this background can increase our understanding of prevailing ideas.

John Lorenz recognized this deficiency, and remedied it with a well-conceived and thoroughly documented review of *Triassic-Jurassic Rift-Basin Sedimentology*. Part One of Lorenz's book chronicles the historical development of concepts and interpretations of the Newark basins and their deposits, focusing largely on the geology of the "Connecticut" Valley. This was a long and involved segment of the history of geology in North America, beginning in the early 1800s with the tentative ideas of such forerunners as Lyell, Silliman, and Maclure, and continuing through to the more sophisticated concepts of today. Part One helps us appreciate the constructive concepts, as well as the mistakes, of others before our time.

Part Two is a synthesis of current information and ideas about the rather similar successions of sedimentary rocks in the numerous Newark basins. Each of seven chapters presents a different lithofacies, ranging from alluvial-fan to eolian and marine-influenced deposits. Examples illustrating the seven facies were selected from appropriate Newark basins, as well as from correlative formations in Morocco, England, and Greenland. Each chapter in Part Two can stand alone as a balanced review of a particular facies, yet through all of them the author has traced several basic themes: the relationships among the varied coeval nonmarine deposits within rift basins; the different methods and techniques involved in collecting and analyzing the pertinent data; and the importance of identifying and interpreting scattered sources of published information. The result illustrates the way that the interplay of established stratigraphic methods and new analytical techniques has led to the present-day reconstructions. It also demonstrates the increasing use of models and of knowledge of modern environments in the interpretation of ancient analogues.

John Lorenz's book is a rather unusual and welcome combination of history of science and geological review. It is informative and readable, reflecting the judgments of an experienced eye. His treatment reinforces the value of both historical perspective and a broad focus on geological problems. Each of the two parts can be perused profitably by itself, yet together they constitute an accessible source of basic information for anyone concerned with Newark research. The contribution is additionally useful because these types of rift basins are now recognized as significant elements in the plate tectonics framework.

I recommend this book especially to all professional sedimentologists interested in understanding why they believe what they do. It will also provide students with an excellent example of how geological ideas develop, and a strong suggestion not to settle for the pursuit of one particular avenue of analysis.

Franklyn B. Van Houten
Princeton University

Preface

The purpose of this book is to illustrate current and historical concepts of sedimentology that have been used within Triassic–Jurassic rift basins along the North Atlantic seaway. The variety of patterns of sedimentation and of sedimentological interpretations of these basin fills, despite the overall similarity of the tectonic regime, emphasizes the complexity of the geological processes that produce the sedimentary record.

In describing aspects of these deposits and how they have been interpreted, this text is intended to serve both as a summary of the Triassic rift-basin strata and as a teaching tool for the methods of studying nonmarine sedimentology. It is not meant to be an examination of the large-scale tectonics of continental rifting, of the stratigraphy and correlation between the numerous basins, or of the many other geological concepts that these basins and their sedimentary deposits might be used to illustrate. The variability in the sedimentology of the different basin fills and the causes for that variability are addressed here, as are the methods by which geological history can be interpreted from such a sedimentary record.

The Late Triassic–Early Jurassic rift-basin sediments were chosen as the subject of this text because they are reasonably well exposed and because they have been studied in considerable detail. Some of these rocks have been studied for as long as the science of geology has been taught in America; yet their interpretations have changed significantly, and a number of controversies still remain. These historical changes and historical differences of opinion illustrate the progress of a number of geological concepts and are instructive in their own right. Viewed together, they suggest that a geologist should remain open-minded, since they show progress to be a process of fits and starts, of multiple and often mutually exclusive hypotheses.

Present-day geologists should find it instructive to examine the evidence—much of which is itself questionable in modern light—that has been offered as irrefutable proof for wrong conclusions. Part I of this book describes how the scientific method has been applied (or misapplied) in the course of almost two centuries of geological study of the North American rift basins. Part II discusses the sedimentological evidence and the present-day interpretations of that evidence that have been developed for various rift-basin deposits on both sides of the Atlantic.

Although there is at present a useful trend toward quantification in geology, nonnumerical methods will always be an important part of this science of myriad variables—as is amply demonstrated in the examination of this suite of complex deposits. At the same time, the geologist must remain aware that artistic interpretation includes many pitfalls, and that several different interpretations of a given incomplete set of data are usually possible.

Geologists, especially field geologists, often cringe under the derision of members of the "hard" numerical sciences. Without detracting from those disciplines, one ought to remember that drawing conclusions from the vast and heterogeneous, yet often incomplete and enigmatic data base provided by the geological record is just as hard (pun intended). Experiments in this field are usually not performable, let alone reproducible. Rather, field geologists are given the superimposed results of one or more of nature's "experiments" and are asked to reproduce the conditions and components of the tests. Gould (1986) has recently written an excellent defense and explanation of the often-misunderstood historical sciences of this type.

The value of exercising caution in the use of closed-system, numerical applications to geological problems is illustrated by the account of Lord Kelvin's calculations of the age of the earth based on rates of cooling (see Burchfield 1975), prior to the knowledge of radioactive decay. Kelvin's conclusions created an oppressive scientific atmosphere, where many geologists and other natural scientists turned their backs on their own qualitative data and crowbarred their ideas to fit Kelvin's terribly short, absolute time scales.

Others, like Huxley, maintained their belief in a long earth history by suggesting that there must have been some unseen flaw in Kelvin's arguments, some as-yet unknown variable or variables. Their viewpoint was bold, unsubstantiated, and right. Despite the use of mathematical techniques, Kelvin and the physicists were, as noted by Eiseley (1958, p. 234) "hopelessly and, it must be added, arrogantly wrong."

Acknowledgments

It is insufficient recompense for their efforts, but I would like to acknowledge and thank the following people, who have given me invaluable and cheerful help. Lorna Bloomberg, Cathy Casper, and Renae Solether kept my correspondence in hand and turned a wild long-hand manuscript into a typed text. Eunice Becker, Pat Chisholm, Linda Erickson, Sharon Gorman, Peggy Poulsen, Nancy Pruett, Connie Souza, and Carmine Ward were able to locate and procure numerous references, many of them obscure, through interlibrary loan. Ron Andree, Merideth Edwards, Arnold Puentes, Steve Scatliffe, and Dan Thompson are responsible for the high quality of the line drawings adapted for use here. John Hubert, Lee Krystinik, and Franklyn Van Houten generously gave considerable time and effort to conscientious and constructive reviews of the manuscript.

Mark Lorenz and Nate Lorenz kept me from losing all perspective and becoming lost in a Triassic time-warp. Ray Doughtwright of Holyoke Water Power generously provided access to outcrops in the Holyoke Dam spillway. Don Schutz spent time in the field recently, guiding me to outcrops from his study of thirty years ago, and Jack Byrnes showed and explained his field area to me back in 1973.

Sandia National Laboratories allowed me the use of technical facilities for the preparation of this manuscript. Finally, I would like to thank the primary authors who provided photos from their work for many of the illustrations, as credited in the figure captions.

PART I

EVOLUTION OF CONCEPTS AND INTERPRETATIONS OF THE HARTFORD AND DEERFIELD BASINS

Introduction

Usefulness of the Historical Perspective

When the towns of Hartford and New Haven were settled in the early 1600s, the settlers knew enough of the rudiments of geology to recognize and make use of the local, low-grade iron ores. Percival (1842) indicates that iron concentrations in some of the basalts were mined and smelted in the East Haven area prior to 1680. Neither a scientific literature nor a scientific concern existed, however, and virtually the only geological fact recorded was that these erratically distributed and very low-grade ores were used until the richer metamorphic iron ores of northwestern Connecticut were discovered and developed.

Although the modern concepts of geology were being developed in Europe near the end of the 1700s, the first scientific descriptions of the rocks of the Hartford basin were not made until the first decades of the 1800s. These early accounts often took the form of travelogues in which localities for mineral collections were listed, together with descriptions of the minerals. Indeed, Benjamin Silliman, who made most of the earliest observations, gave his early geology lectures at Yale University on the subjects of mineralogy and chemistry—disciplines that covered most of what was then understood of and deemed important to the science.

In the 175 years since Silliman's first observations, the science of geology, the subdiscipline of sedimentology, and the knowledge and interpretations of the rocks of the Hartford basin have undergone tremendous changes. The paths of these advances, lumped together under the term *progress,* might be described as zigzagging erratically upward with time—sometimes with multiple pathways—between the walls of a triangle: the triangle's walls would represent the boundaries of the ideas considered tenable at a given time. From our present-day perspective, we can see that we must be approaching

3

the triangle's apex (the geologically correct interpretations), but we cannot see how close we are to that pinnacle. Certainly we have come a long way, considering the width of the triangle's base; but each generation of geologists has believed that it was close to or at the top, only to have its ideas superseded by those of the next generation with its increased stock of data and more refined geological models.

Part I of this book retraces the zigzag course of geological knowledge in the Hartford basin from the early nineteenth century to the present and examines the arguments and evidence used by various geologists in support of their conclusions. It also examines the interaction between the growing science of geology, and the evidence and interpretations produced from one assemblage of deposits. Much of this interaction was one-sided, as geological models from all over the world were developed and applied to the Hartford basin. But reversals occurred, too, especially during the 1800s when the scientific center of gravity in this country was located in the academic communities of the New England universities—several of which were built on (and often used building stone quarried from) the deposits discussed here. As noted by Sanders (1974, p. 15), "Perhaps more than with any other single large suite of rocks, interpretations of the Newark strata have been closely controlled by the status of ideas prevailing within the fabric of geology."

It is difficult to separate interpretations of the Hartford basin from ideas developed synergistically in studies of other North American rift-basin deposits, and a fair amount of this interaction will be evident. As much as possible, however, this study will focus on the Hartford basin.

Today's sedimentological studies do not refer to most of the older works cited here because the older works contain antiquated and often incorrect ideas; and when they are referenced, such works are usually dismissed as being of historical interest only. The appended *only* is unnecessary, and implies that the history is irrelevant. Any scientist can benefit from knowing how the fundamental concepts of her or his discipline came to their present status, and what mental conflicts and debates were necessary to bring about their recognition and acceptance.

The 1800s and early 1900s were not the dark ages of geology. Krynine (1950) noted that over 250 titles on the rocks of the Hartford basin had been published, and that well over 1,200 papers dealing with the Triassic–Jurassic of eastern North America existed. Most of these early papers did relatively well in interpreting the rocks, considering that the authors were saddled with the Bay of Fundy as the acceptable modern analogue. They also had no inkling that these rift basins were situated in the interior of a supercontinent that was in the process of breaking up—rather than being located, as they are presently, marginal to the ocean—or that the deposits were nonmarine and predominantly subaerial.

Figure I-1. Example of poor exposure in natural outcrops. The Holyoke Basalt flow forms an erosion-resistant ridge, but vegetation covers most of the associated sedimentary rocks. The artificial cut at the bottom of this photograph is part of a basalt quarry. (Photo by the author)

It is interesting to read papers from this era. The phraseology, spellings, and format are of a style we would consider antiquated, but they were up to date when written. The language sometimes has a baroque style: remnants of an era when, to quote Eiseley (1958, p. 18), the purpose of science was "to name and marvel." These papers, containing many precursors to present linguistic conventions and terminology, illustrate the utility of flexibility in language.* The papers are also poorly referenced, noting an author and his ideas but often not the year and journal. The geological community was so small that other geologists were usually aware of the work being referenced.

Although the early geologists lived close to the deposits they were studying, their job was not an easy one. Outcrops of the rocks were few and far between owing to the temperate climate, which produced rapid erosion of the deposits and a dense growth of vegetation (figure I-1). This condition is exacerbated by the blanket of Pleistocene glacial drift so well known to (and cursed by) New England farmers, who continue to pull new crops of glacial cobbles from their fields every spring.

*It would benefit those who currently want to purify and codify English, scientific or otherwise, to read some of this early literature. Who will want to write 1980s-style English 200 years from now? Only dead languages such as Latin offer unchanging vocabulary, conventions, and definitions.

These conditions in turn obscured the postdepositional structural complexities of the basin, which were not fully recognized until the last decades
of the 1800s. The basin's structure is not complex by mountain-belt standards, but until the structural overprints could be subtracted from the deposits, the stratigraphic correlations and facies relationships that are defined
by the three intercalated time-line, basalt, lava-flow units could not be properly sorted out, and detailed basin analysis was impossible.

As the countryside was developed, many of the detailed geological observations began to come from artificial outcrops such as road cuts and quarries. Because of the apparent absence of coal or oil in the formation, few
wells have been drilled, and subsurface data are virtually nonexistent.
Recently, renewed speculation on the hydrocarbon potential of these rocks
has occurred, and we may soon have some well-documented drillholes into
the formations. Except as it relates to building stone, however, the economic
history of these rocks has been rather dismal.

Although this book addresses primarily the sedimentological aspects of
the rocks that fill the rift basins, the scope of interest in Part I is slightly
broader, inasmuch as developments and interpretations of sedimentology
were built on the older subdisciplines of geology. As much as possible,
however, the theme in Part I remains the sedimentological developments
and interpretations.

Much good sedimentological study is now going on in these basins. The
Connecticut and New Jersey areas in particular are the focus of present-day
studies by the U.S. Geological Survey and by geologists at several universities
and oil companies. Some twenty abstracts on different aspects of the eastern
North America rift basins were printed in 1985 by the Geological Society of
America and by the American Association of Petroleum Geologists, each
abstract representing an ongoing study.

Review of the Newark Supergroup in the Hartford and Deerfield Basins

The brief description of the sedimentary rocks and their currently accepted
interpretations that follows is intended to establish a reference framework
for the subsequent historical account. The papers published in the late
1970s and early 1980s by Hubert and his associates/students are the principal
basis for this description, supplemented by other recent work. (These papers
will be referenced fully in later sections of the book.)

Stratigraphy

The Newark Supergroup, named for rift-basin deposits in the Newark-
Gettysburg basin of New Jersey and Pennsylvania, is a series of correlative

deposits infilling rift basins along the eastern continental margin of North America (Van Houten 1977; Olsen 1978).

Although Olsen (1978) and Froelich and Olsen (1984) have incorporated the strata that occur within the Hartford basin into the Newark Supergroup nomenclature of rift-basin deposits, at present there is no group name that indicates exclusively the formations of the Hartford basin (Luttrell 1985). Consequently, there is no formal way to refer to them other than by such awkward phrases as "the Newark Supergroup of the Hartford basin," or "the Hartford basin sedimentary fill." In the absence of a simple and authoritative name, these formations will be informally and more conveniently referred to here as "the Hartford group." This follows the general pattern of naming the basin fill after the basin (as with Culpeper Group and Fundy Group), and stratigraphers can formalize the terminology if it is warranted. Such a step eliminates the Meriden Group (Lehman 1959), formerly the Meriden Formation (Krynine 1950), but this group name has been of limited use in sedimentological discussions.

The Hartford group in Connecticut is divided into four sedimentary formations, separated by three named basalt lava-flow units (figures I-2 and I-3). These formation names have not been extended in toto north into

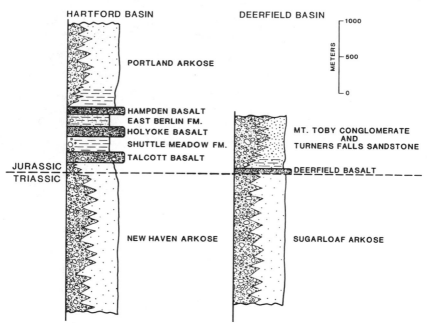

Figure I-2. Stratigraphic column of the rocks in the Hartford and Deerfield basins, Connecticut and Massachusetts. (After Hubert et al. 1978, Figure 3. Reproduced with the permission of the University of Massachusetts, Dept. of Geology and Geography)

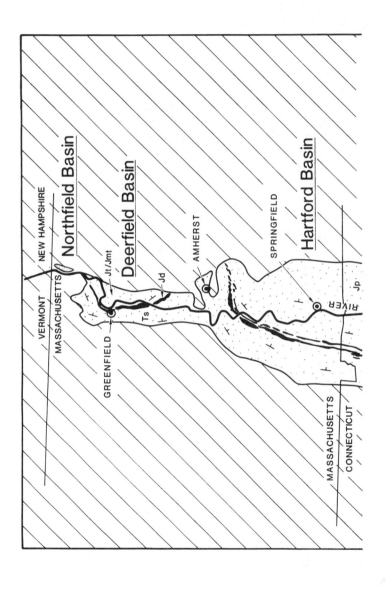

Northfield Basin

Deerfield Basin

Hartford Basin

NEW HAMPSHIRE

VERMONT

MASSACHUSETTS

GREENFIELD

Jt/Jmt

Jd

Ts

AMHERST

SPRINGFIELD

RIVER

Jp

MASSACHUSETTS

CONNECTICUT

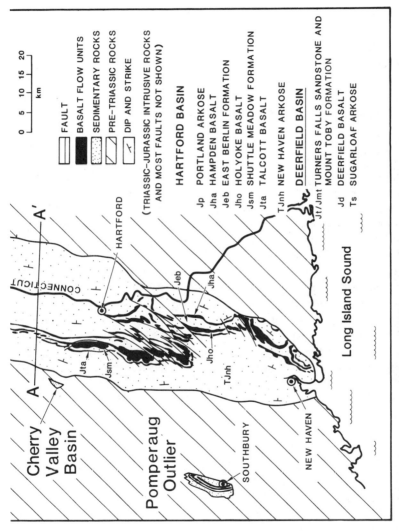

Figure I-3. Generalized geology map of the Hartford and Deerfield basins. (Adapted from Rogers 1985; Wheeler 1937; and Zen, ed. 1983, with the permission of the U.S. Geological Survey, the Connecticut Geological and Natural History Survey, and the Massachusetts Dept. of Public Works)

Massachusetts, where the formations were separately defined on the basis of common lithologic character rather than on their stratigraphic relationship to the basalt flows.

These deposits are Late Triassic to Early Jurassic (and possibly Middle Jurassic) in age. The Hartford group represents the younger end of the Middle Triassic to Early Jurassic spectrum of rift-basin deposits along the east coast of North America. As reported by Hubert et al. (1978), the approximately 3,700 meters of sedimentary and igneous rocks preserved in the basin represent about 36 million years of Late Triassic to Early Jurassic time, from about 201 million to 165 million years before the present.

Basin Structure

The present structure of the Hartford basin is essentially an asymmetric half-graben, with beds dipping 15 to 20 degrees to the east toward a major, probably step-faulted, eastern border fault (figure I-4). The strata either lap unconformably up onto the basement rocks at the opposite western margin of the basin or, locally, are faulted against them. Numerous parallel, postdepositional, normal and strike-slip faults transect the basin at angles oblique to the basin axis and the eastern border fault. The eastern border fault is also normal (probably listric) and currently dips about 55 degrees to the west. This configuration is an exaggeration of the syndepositional structure, which is considered to have been fault-bounded on the east, with rotation of the basement rocks during subsidence such that subsidence was much greater near the eastern border than in the western areas.

The current structure suggests that the entire basin was postdepositionally

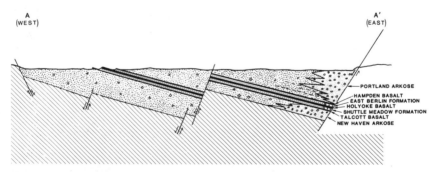

Figure I-4. Schematic east–west cross section of the Hartford basin. See figure I-3 for location. (After Krynine 1950; Sanders 1968; and Rodgers 1985, with the permission of the U.S. Geological Survey and the Connecticut Geological and Natural History Survey)

rotated, down on the east. Postdepositional erosion also occurred, so that the present basin-fill sedimentary rocks represent an unknown, but probably significant, percentage of the original basin fill. Other small remnants of correlative sediments are present in two areas, preserved as outliers faulted into the metamorphic basement west of the central basin (figure I-3). These outliers record a preerosional extension of the same deposits farther to the west, although the thinness of the formations in the outliers indicates that the western margin of the basin was probably well east of the New York border. A third, recently discovered outlier of correlative deposits, located north of Boston, east of the eastern border fault (Kaye 1983), was probably not originally connected to the central basin.

The Hartford basin is partially separated from the Deerfield basin to the north by a structural basement high (the Amherst inlier). This basement block was apparently high during deposition—although some of the Triassic-Jurassic strata are continuous over it—and it was not a sediment source (Chandler 1978). The much smaller Northfield basin lies slightly to the north (figure I-3), and is probably an erosionally separated northern extension of the Deerfield basin.

Sedimentology/Depositional Environments

Hubert et al. (1978) have summarized many of the currently accepted interpretations of depositional environments for the different formations found in the Hartford basin, and this description follows their summary for the most part. Much work, however, is currently being done in refining these various models.

The basal formation is the New Haven Arkose, estimated to be about 2,000 meters thick. On the eastern margin, against the eastern border fault, this formation consists primarily of red arkosic sandstones and conglomerates derived from igneous and metamorphic source terranes directly to the east. The conglomerate clasts are often coarse and angular, reflecting the proximity of their source and the high relief prevalent during the deposition of these fluvial and alluvial-fan deposits.

The coarse deposits grade rapidly westward (usually within a few kilometers) into facies-equivalent red arkosic sandstones and mudstones (figure I-4) deposited in braided-stream environments. Streams flowed generally to the southwest. The climate was probably semiarid and seasonal, and carbonate caliche profiles were commonly developed on the interfluves and abandoned fluvial sediments. Farther west, meandering fluvial systems may be recorded in some of the equivalent deposits found in the Pomperaug outlier.

This episode of deposition was interrupted by the outpouring of a

65-meter-thick series of basaltic lava flows that formed the Talcott Basalt. This sequence is the only one of the three basalt formations that exhibits extensive pillowing. It also displays some of the rare evidence of explosive volcanic activity in the basin.

Following this volcanic episode, and possibly after an erosional hiatus, the 100-meter-thick Shuttle Meadow Formation was deposited. The muddy and locally calcareous sediments of this formation extend the width of the basin, except for local conglomeratic deposits adjacent to the eastern border fault. Deposited in a series of playa and perennial lacustrine environments, possibly during a climatic trend toward increased precipitation, this formation contains both fossil fish and the earliest record of common terrestrial vertebrates found in the basin (in the form of footprints in the mudstones).

Renewed volcanic activity produced the thickest (100 meters) and most extensive basaltic unit in the basin, the Holyoke Basalt. Relatively uniform flow thicknesses and their widespread nature suggest a highly viscous series of flows.

After this period of volcanic activity, lacustrine and playa conditions returned to the basin, and the mudstones, sandstones, and local limestones of the 170-meter-thick East Berlin Formation were deposited in conditions similar to those of the Shuttle Meadow Formation. These strata also contain fossil fish and footprints.

A third volcanic episode produced the 60-meter-thick series of basalt flows of the Hampden Basalt, after which tectonic and climatic conditions slowly returned to those under which the New Haven Arkose had been deposited. At least 1,200 meters of conglomeratic, sandy and muddy deposits accumulated in alluvial-fan and (grading westward) braided stream environments, to form the Portland Arkose. Lacustrine black mudstones are interbedded with these alluvial-fan and fluvial deposits, primarily in the lower half of the sequence. Above this formation, the sedimentary record is lost due to postdepositional erosion.

Similar lithologies and environments are recorded farther north in the Deerfield basin in Massachusetts, although the stratigraphic sequence is divided differently. The Deerfield basin is smaller, and there are suggestions that its lakes and border-fault alluvial fans may have been developed at different times than those to the south.

Tectonics

The strata of the Hartford group overlie an erosional unconformity (figure I-5). They were deposited in a basin that was bounded on the east by a series

Figure I-5. Triassic conglomerates of the basal New Haven Arkose overlying an erosional unconformity and pre-Triassic metamorphic rocks. (Photo by the author)

of syndepositional north-south normal or listric faults connected by northeast trending faults. Subsidence and cumulative deposition were more rapid at this eastern margin than in the western part of the basin. Subsidiary border-parallel faults may also have been present, giving rise to step faulting along the border, but they are now buried and must be inferred from indirect evidence such as gravity modeling and seismic profiling.

The nature of the present western edge of the basin is more ambiguous. In places, the sediments lap up onto the basement rocks; elsewhere, they are found in faulted contact. This edge of the basin has been variously interpreted as either the faulted mirror image of the eastern faulted margin or as the erosional edge of a basin that originally extended all the way to the Hudson River. Currently, it is interpreted as the eroded margin of a thinning westward half-graben, modified by postdepositional faulting and erosion.

The regional tectonic regime during Late Triassic–Early Jurassic time was one of extension during the breakup of Laurasia. Some authors have suggested a component of strike-slip motion for the opening of the rift basins that formed on this extending continent, but the question is still open. In the case of the Hartford basin, as elsewhere, the preexisting structures of the basement probably controlled the location and orientation of most of the basin-margin faults—and the axis of the resulting basin—as much as did the prevailing tectonic stresses (Swanson 1986).

Postdepositional tectonics accentuated the easterly tilt of the beds and created numerous oblique cross faults within the basin. This, together with postdepositional erosion, added complexities to the pattern of conglomerate-to-mudstone east-west facies changes and to the fluvial-lacustrine-fluvial vertical changes in depositional environments.

1 / Early 1800s

Silliman and Maclure: Earliest Observations, European Roots

Beginnings of American Geology

The rocks of the Hartford group, obscured by vegetative and glacial cover, had yet to be described and interpreted when Benjamin Silliman, age twenty-two, was appointed to the newly created chair of chemistry and natural science at Yale University in 1802. This was the first science position other than mathematics to be established in an American university; American education had long been dominated by the church and its opposition to natural science. "The creation at Yale in 1802 of a professorship of chemistry and natural science was not only unprecedented but daring" (Ogburn 1968, p. 38). Silliman was a product of his times, however, and (Ogburn continues) "as Thomas Huxley later put it, he wrote 'with one eye on fact and the other on Genesis.'" This is understandable given that Silliman had to prepare himself for his position from scratch (his training was as a lawyer) and that the intellectual atmosphere required him to do his studying surreptitiously (Merrill 1906).

After lecturing for several years, Silliman was given funds to purchase laboratory equipment and a library, and he headed for Europe and the contemporaneous centers of scientific progress in order to do so. As chronicled by Fenton and Fenton (1952), Silliman was able to attend lectures at the university in Edinburgh, Scotland, which was becoming deeply embroiled in the Wernerian versus Huttonian debates on the origin of rocks, especially basalt. In essence, Werner held that many rocks were attributable to universal and sequential chemical precipitation from a worldwide, subaqueous episode of biblical proportions. Hutton, the originator of uniformitarianism, believed in the igneous origin of basalts. Werner taught that clastic rocks had been deposited by currents produced during the withdrawal of his universal ocean, while Hutton ascribed them to presently observable processes. Silliman took these conflicting ideas back to Connecticut and began to make the first scientific observations on the local geology.

Silliman's first article on Connecticut geology was published in 1810. It made use of the Huttonian concepts of modern processes to suggest that New Haven was built on a plain of sediments—now known to be primarily glacial outwash (Skinner and Rodgers 1985)—derived by erosion from the surrounding highlands. Silliman also described the local Mesozoic basaltic intrusions, and although he obviously leaned toward a Huttonian/plutonist interpretation of these as well, he presented both theories and refrained from drawing specific conclusions.

In this and in subsequent publications until as late as 1830, Silliman was always careful to hedge his bets, leaving open an opportunity for recantation should his arguments in favor of a volcanic origin for the basalts ever be disproved. He did so despite the fact that he could demonstrate basal contact metamorphism and vesicular flow tops in many of his outcrops.

These descriptions and interpretations of the basalts dominated the attention of Silliman and most of the other early geologists who attempted to understand the geology of central Connecticut. The basalts were often called *whinstones* (after the basaltic Whin Sill in England), *greenstones*, or—most commonly—*trap rocks* (after the Swedish word *trappa*, for *stairs*, which successive flows resembled in places in Europe). The term *trap rock* is still commonly used in Connecticut to describe the road metal (base material over which tar or concrete is poured) that is quarried from these flows. In fact, the basalts were the most conspicuous of the lithologies of central Connecticut, being more resistant to erosion and better exposed; thus, many of the earliest references to the Hartford and Pomperaug basins were to the "basins of greenstone" (Silliman 1818, 1820), despite the relatively small proportion of the basin fill that is basalt.

Correlations with European Geology

Maclure (1809) had recognized the sedimentary nature of the greater part of the basin fill. He had mapped both central Connecticut and the Pomperaug outlier as Old Red Sandstone of the "Secondary" formation (figure 1-1), using the Wernerian scheme of rock classification and correlations with the lithologically similar Devonian strata of Europe. Maclure was a uniformitarian in philosophy however, and he reserved judgment as to genetic interpretations of the rocks. His use of Wernerian nomenclature signified only that no better alternative existed at the time.

In fact, Maclure (1824) published several notes on observations of multiple intercalations of sediments with basalts—an impossibility according to the Wernerian doctrine that given rock types followed a given worldwide

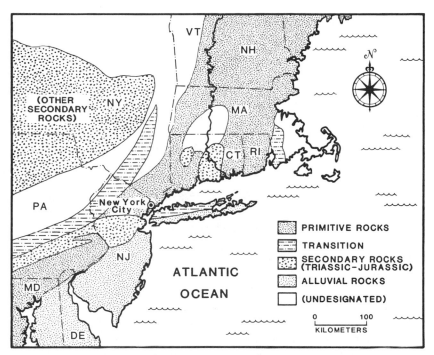

Figure 1-1. 1809: First geological map that distinguished the rocks of the Hartford group (after Maclure 1809). Note that the Pomperaug outlier had been recognized. Original map was hand colored; without lines drawn between the colors, they often overlapped.

stratigraphic sequence. He also philosophized that, while Werner's scheme might be incorrect, there probably was some definitive order to the global succession of rocks, as yet unknown.

This early correlation with the Devonian, based solely on lithologic similarities, led to fruitless search in Connecticut for the Carboniferous coal-bearing strata that should have overlain the "Devonian" redbeds, *if* the age correlation was correct and *if* the analogy with the European succession was applicable. The economic geologists were tempted onward in the search for coal in Connecticut by the discovery of coal-bearing strata in equivalent red rift-basin strata in Virginia, and by thin carbonaceous streaks present in some horizons in the Hartford group, often reported as coal. Although early exploration was carried out (Silliman 1821), no true coals were ever found.

The term *Triassic* was not proposed until 1834, and then its type deposits were the entirely different, widespread sandstones and evaporitic mudstones of the Germanic, cratonic sequence. Therefore the acceptance of the American rift-basin deposits as age equivalents to these deposits did not take place until the mid-1800s and the discovery of fossiliferous zones in both sequences.

Origin of Connecticut's Basalts

Meanwhile, Silliman continued to make observations on the geology of the entire state of Connecticut, concentrating primarily on mineralogical descriptions and outcrop locations, usually with an eye to economic geology rather than to interpretations. In 1818, he established the *American Journal of Science,* which after a precarious start soon narrowed its scope primarily to geology. Silliman's travelogues continued to describe the basalts and the types of secondary mineralization commonly found in the vesicles, until in 1822 Thomas Cooper sent a manuscript to him (in care of the *American Journal of Science*) that cogently argued the case for the igneous origin of basalts.

Silliman published the article with a footnote indicating that the conclusions as to the igneous origin of the Connecticut basalts were Cooper's alone, and that he (Silliman) was still reserving judgment. He then fell silent on the subject—in print, at least—until 1830. At that point, he published a paper of his own that borrowed heavily from Cooper's arguments, and mixed them with a curious blend of science and emotional appeals to the reader. After noting the general characteristics of lavas (much as Cooper had), Silliman described an extensive quarry cut near Hartford. In this cut, he noted contact metamorphism and the gradation within the basalt from crystalline basal to vesicular upper zones. Silliman even made the analogy with modern lava flows and with contemporaneous experiments showing that the degree of crystallinity is a function of the rate of cooling. Then he followed up his conclusions as to the basalts' igneous origin by stating that he had brought several visitors to the quarry, all of whom were of "very indubitable character," and all of whom agreed with him, "whether interested in such subjects or informed in geological facts and theories or not" (Silliman 1830, p. 127).

At the end of the paper, though, Silliman was still hesitant to make a blanket statement concerning the igneous origin of basalts; he even suggested that there might eventually be sufficient evidence to "admit the aqueous origin of trap in some cases" (Silliman 1830, p. 131). Such was the persuasiveness of the Wernerian arguments he had heard some twenty-five years previously, or perhaps such were the constraints of the intellectual climate.

Origin of Concepts of Subaqueous Sedimentation

Of greater interest to the sedimentologist than Silliman's arguments on the origin of basalts are his observations in this same 1830 paper on the conditions of deposition of some of the associated sedimentary rocks. Silliman began with the Wernerian premise that there was an "ancient ocean which enveloped the globe before it attained its habitable condition" (Silliman 1830, p. 129). According to Barrell (1906), this view originated with (or at least was strengthened by) the fact that the early geologists made their observations in Europe and eastern North America—areas of continental erosion. All deposition in those areas seemed to be taking place only in the adjacent seas. Biblical concepts of Noah's flood undoubtedly were also influential.

Silliman noted two sedimentary features that supported his preconceptions that the sediments were laid down underwater. The first was an instance of bedding-plane ripple marks, recognized (but not named) as having been produced by "undulating water." Associated features, described as "irregular lines, running upon their [the rock] surfaces, like large veins just under the skin of an animal," (Silliman 1830, p. 123-24) were probably an example of desiccation cracks. They were noted but not interpreted.

The second feature presented by Silliman as evidence of subaqueous deposition is an interesting example of evidence incorrectly interpreted, although the misinterpretation was used to support a generally correct (subaqueous) hypothesis. Silliman noted that in quarrying conglomerates, "cavities in the rock which were entirely secluded from the atmosphere were often found full of wet clay," (Silliman 1830, p. 130), and he concluded that these supposed remnants of depositional waters proved a marine origin.

The concept of an ancient global sea, derived from biblical and Wernerian concepts and tenuously (by recent standards) supported by evidence of subaqueous deposition in the rocks, was to define the limits of the interpretations of depositional environments of the Hartford group for the next century. Never mind that the evidence dictated neither fresh nor saline water, that some of the evidence was misinterpreted, and that, as geological concepts evolved, modern nonmarine environments were recognized. Until the earliest 1900s, despite all the evidence for nonmarine deposition subsequently discovered in the Hartford basin (including plants and a nonmarine fauna of vertebrates, invertebrates, and footprints), most theories of depositional environments included some form of marine influence. Many geologists incorporated it into the theme of their papers as an afterthought, a ceremonial bow to conventional wisdom; others inserted the theory with a crowbar. In any event, it was an idea that died a lingering death. Benjamin Silliman, its American foster-father, has been remembered over time primarily as a teacher and as the founder of the *American Journal of Science*.

Hitchcock: Ichnology and Early Sedimentology

Detailed Mapping

Before eventually becoming its third president, Edward Hitchcock was professor of natural theology and geology at Amherst College. He was also one of the first state geologists associated with the era of state surveys, a period from 1830 to 1880 during which individual states were funding geological assessments and scientific descriptions of their lands (Merrill 1906).

Hitchcock had published a geological map (figure 1-2) and a mineralogi-

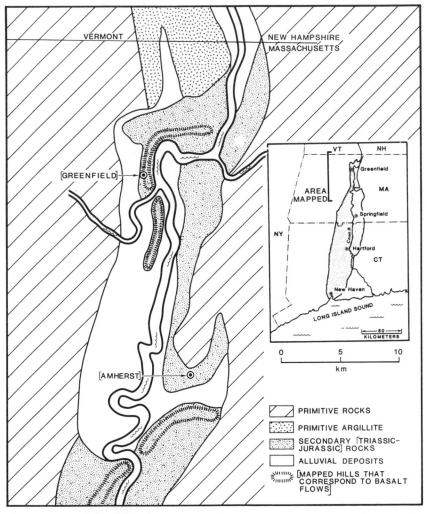

Figure 1-2. 1818: Geological map of the Deerfield and northern Hartford basins (after Hitchcock 1818). Note that no distinction was made between the basalt ridges and the sedimentary units, even though the ridges were delineated as geomorphic features. At this time, both the basalts and the sedimentary rocks could theoretically be attributable to the same depositional process. The cross section published with this map depicts "greenstone" (basalt) units as essentially vertical beds.

cal travelogue of the Hartford basin in Massachusetts in 1818, but his first substantial work appeared in 1823 and 1824 in a series of papers entitled "A sketch of the geology, mineralogy, and scenery of the regions contiguous to the River Connecticut." This series of papers interpreted the basalts as igneous, and followed Maclure's lead in ascribing most of the strata to the Old Red Sandstone of Werner's Secondary formation. Hitchcock noted, however, that part of the sedimentary sequence (the present Shuttle Meadow and East Berlin Formations) could be lithologically differentiated from the rest. Since these rocks were darker and contained both fish fossils and (locally) thin carbonaceous streaks, he logically ascribed them to the post-Devonian "Coal Formation" or Carboniferous of Europe.

Hitchcock accompanied his 1823-24 papers with an expanded and improved geological map of the rocks filling the Hartford basin. This map (figure 1-3) is revelatory primarily for what it does not show, including faults and many basalt outcrops. Mapping an area this large (some 175 by 65 kilometers) on horseback, without base maps of any value and with abundant cover, was a major and difficult undertaking. The resulting map failed to reveal important patterns of the outcrops by which the rocks might have been interpreted, but it was an important start.

Hitchcock wrote his early papers with a basically Wernerian outlook. In several passages, he was obviously puzzled to find himself describing "Old Red Sandstone" as occurring both below and above the "Coal Formation," and elsewhere, intercalations of basalts at several horizons within the "Coal Formation." His preconceptions dictated that there should be a single horizon for each lithology, and that these should be arranged in an orderly, predictable succession. Hitchcock even tried to diagram a basin wherein the dip of the individual beds was uniform across the basin while, at the same time, the dip of the formations was not (figure 1-4)—as a means of keeping the "Old Red Sandstone" in its proper position despite apparently contrary surface data. He later decided the redbeds above the "Coal Formation" might be correlative with the New Red Sandstone (Permian–Triassic) of England.

As Hitchcock's understanding of the basin improved, and as the general body of geological knowledge expanded, Hitchcock's map changed. His 1844 geological map of Massachusetts (figure 1-5) showed the same basic patterns as his 1823 map, but the formations previously mapped as "Old Red Sandstone" and "Coal Formation" were now combined and mapped as "New Red Sandstone." Each map still portrayed a combination of surficial and bedrock geology, but by 1844 Hitchcock had recognized that some of the igneous rocks (known today as the Hitchcock Volcanics) were tuffaceous, and distinct from the basalt lava flows. He envisioned these tuffs as having "poured forth at the bottom of the ocean" (Hitchcock 1844, p. 7).

Hitchcock was the first to describe the fossil bones and fish impressions

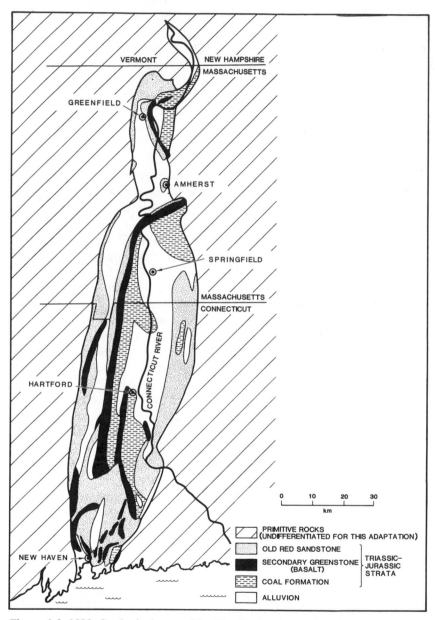

Figure 1-3. 1823: Geological map of the Hartford and Deerfield basins (after Hitchcock 1823–24). Original map was hand colored and used different shades of blue for the same rock types. Although no distinction was made between the intrusive and extrusive basalts, the basalts were mapped as separate units from the sedimentary strata. Few of the stratigraphic patterns of the basin are apparent, however, due to inaccuracies of both the base map and the geological mapping.

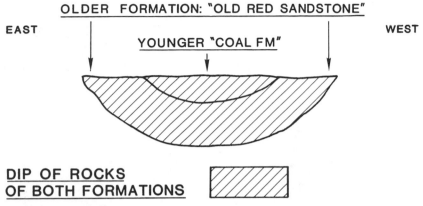

DIP OF ROCKS OF BOTH FORMATIONS

Figure 1-4. 1823: Hitchcock's schematic cross section—an attempt to rectify the observations of constant structural dip and repeated lithologic sequences with Wernerian dogma (after Hitchcock 1823, p. 42). Hitchcock had convinced himself that the obvious dip of the beds did not reflect dip of the formation. Thus, "the strata of both these rocks have their dip in such a direction as to lead one, at first, to conclude that this old red sandstone lies [both below and] *above* the coal formation. It does not [necessarily] follow, however . . . and it was not till I had traversed it a third time that I felt entirely satisfied" (Hitchcock 1823, p. 41–42). This concept may have been derived from contemporaneous theories of originally inclined dips of bedding.

common in parts of the Newark Supergroup. Many of his early papers were the sort of descriptive travelogues common in papers of that era; he wrote an entire paper (1824) describing the scenery along the Connecticut River, and he related the story (1828) of a farmer who tried to split a piece of fossil wood for firewood, and upon ruining his axe, "flew into a passion, and fell to breaking this fine petrification to pieces" (Hitchcock 1828, p. 228). Hitchcock was an astute observer, however, and was adaptable to the changes in scientific venue. During the flood of the Connecticut River in 1840, Hitchcock evidently observed the formation of a meander cutoff and oxbow lake along the river (Brophy et al. 1967, quoting a newspaper account). This must have reinforced his increasingly uniformitarian leanings, since such uniformitarian fluvial processes had been described in Lyell's *Principles of Geology* only ten years earlier. Hitchcock was to become adept at making deductive interpretations in sedimentology, in paleontology, and in ichnology, a field he played a strong role in developing.

Ichnology

Hitchcock made important and controversial contributions to the subject of ichnology—and probably wished he had not. Lull (1915) credited him with

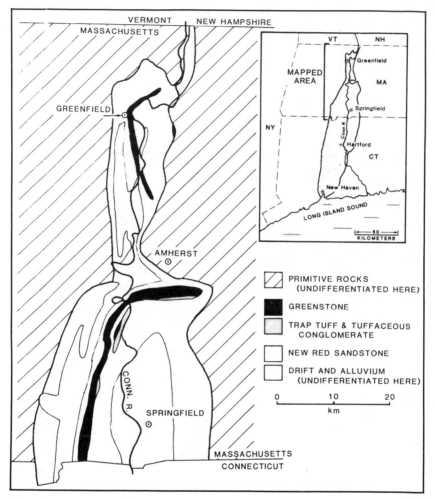

Figure 1-5. 1844: The Hartford basin portion of Hitchcock's 1844 geological map of Massachusetts. The map was again hand colored, which as Hitchcock noted (in the era before electric lighting) made distinctions of colors difficult, "especially in the evening." The colors do not overlap, however, and a number scheme was added for clarity. All sedimentary rocks in the basin were now ascribed to the New Red Sandstone (correlative with that of Europe), and distinctions were made between glacial drift and alluvium, and between the basaltic lava flows ("greenstone") and tuffaceous igneous rocks.

being the father of ichnology, and Krynine (1950, p. 12) noted that he made ichnology "a full-fledged branch of paleontology."

What Hitchcock did, starting in 1836, was to recognize the abundant vertebrate and invertebrate tracks present in parts of the Hartford group for what they were, rather than as evidence of "Noah's raven" as they had been popularly described. On this basis, he began to describe scientifically the tracks and the animals that made them. He was at a loss for exact modern analogues, however. Buckland had not described the first dinosaurs in print until 1824, and Owen would not erect the class Dinosauria until 1841. Furthermore, because of fragmentary evidence and by analogy to large modern animals, all dinosaurs were originally reconstructed as quadrupeds. Hitchcock's tracks were almost exclusively of bipeds. Because the tracks were very much like those of modern birds (down to the fabric of the skin in the better preserved specimens), Hitchcock interpreted his fossils as bird tracks despite the size differences. The discovery soon after of fossil bones of *Dinornis*, a 2- to 3-meter-tall bird, served to strengthen his convictions, as well as to persuade many others of the validity of his interpretations.

Hitchcock began to collect the tracks (they numbered over 20,000 at his death and were housed in a special museum at Amherst College) and to catalog and classify the different types. He coined the terms *ornithichnites* ("stoney bird tracks") and *ornithichnology* (the study of same) for his specimens. Although he and others generalized and shortened the terms to *ichnites* and *ichnology* in later studies, he remained convinced that they were made by birds—despite serious doubts by many of his contemporaries, especially as discoveries of dinosaurs became commonplace. The discovery of *Archaeopteryx* just before Hitchcock's death only served to confirm his views.

In light of present theories of the relationship between dinosaurs and birds, Hitchcock's dogged persistence in his interpretations is laudable and might be perceived as a foresighted theory that was eventually discounted for the wrong reasons. He recognized that the linear pattern of successive tracks along trails indicated animals with an upright bipedal stance, whereas the then-current concepts called for dinosaurs with a reptilian sprawl. He also pointed out that the number and arrangement of toes and claws in the tracks were similar to the number and arrangement of these parts in birds. He made preliminary calculations showing that the ratio of foot size to stride length was that of bipedal, birdlike animals, not that of quadrupeds.

In essence, Hitchcock had approached the question of the affinity of the animals from a direction opposite the one taken by conventional paleontology, studying traces of the animals' behavior rather than their bones. His bird theories died with him, however, as later paleontologists ascribed his tracks and the related bones to "saurian reptiles" and suppressed the bird hypothesis. Modern ichnology, concerned primarily with more environmentally diagnostic marine trace fossils, has made little use of his detailed observations.

In Hitchcock's own time, the fame the tracks earned him was coveted by James Deane, M.D., who had apparently first brought the footprints to Hitchcock's attention but who himself was unqualified to study them. The subsequent controversy over both interpretation and original discovery led in 1840 to the appointment of a committee by the American Association of Geologists and Naturalists to investigate the claims. Hitchcock's interpretations (and obliquely, his claims to priority) were fully supported (Rogers et al. 1841), but the incident left a bitter taste in his mouth, and Hitchcock devoted much space in subsequent publications to a continuing defense of his position.

The animal tracks were cited as evidence for several other theories. Their impression into the rocks normal to the bedding planes was frequently offered as proof of the originally horizontal angle of deposition, although several examples showing that an animal had slipped were also found and were used to support the idea that the present structural dip of the strata was in fact a primary depositional dip. Rogers (1856), one of the few supporters of primary depositional dip, also used the limited diversity of species represented by the tracks as evidence for the theory of progressive development—cycles of catastrophic extinction and creation, attended by the "progressive elevation" of each successive assemblage of life forms. Rogers used these data to argue against Lyell's early concepts of uniformitarianism of life forms.

Early Sedimentology: Marine Models

Hitchcock's recognition of nonmarine animal tracks required a certain amount of rethinking on his part. He had originally concurred with the conventional interpretations of the rocks as having been formed in marine depositional environments; indeed, he (like others) continued to carry this mental baggage with him throughout his career, modifying it until it meshed with his perceptions of the evidence. In 1841, he suggested that, because the rocks were once soft sediments, the birds that made the tracks in them must have been long-legged waders, and therefore they had inhabited the muddy shores of "estuaries, streams, and lakes." The present geographic position of the Hartford basin, apparently open into Long Island Sound, could only have reinforced the concept of estuarine depositional conditions, though no one ever bothered to spell it out.

Hitchcock (1841) explained the nonmarine plant remains that he found as having come from outside the basin, and having "drifted into their present position by current" (p. 450). He also noted raindrop impressions, even noting which way the wind had been blowing during the shower. Because neither the raindrop impressions nor the tracks had been washed away, he speculated that there must have been wet and dry seasons such that the water levels of the "estuary or lake" fluctuated, preserving the impressions at times of drying under "the heat of a powerful sun" (p. 512). In the same paper (1841), however, Hitchcock speculated that the water level fluctuations were probably the result of tidal action.

In this way, without being rigorously definitive or providing an incontrovertible proof, Hitchcock attached more circumstantial evidence to the model of marine (now modified to include estuarine) environments of deposition, and this became ingrained as the conventional wisdom of the early 1800s.

Hitchcock also speculated on the pre-Darwinian concepts of evolution, being an adherent to the theory of progressive extinctions despite his position as (in part) a professor of natural theology. He collected his lectures into a volume called *The Religion of Geology* (1851), in which he discussed such things as absolute time (based on sediment transport rates in the Mississippi River and the volume of its modern delta). He was firmly committed to his religion, believing that the purpose of science was to find proofs for religion, but he also believed in a liberal interpretation of the Bible that would give present-day fundamentalists apoplexy.

This interplay of religion and geology was a popular study for some scientists during this era. Silliman, for example, wrote at length on "The Consistency of Geology with Sacred History" as an appendix to Bakewell's *Introduction to Geology* (Silliman 1833). Others considered such exercises inappropriate. Some complained that having Silliman's appendix in a geology text confused two studies that were usually taught as separate disciplines (Merrill 1924, p. 154), while others noted that the rapid progress in geology necessitated new reconciliations with scripture with unsettling frequency (Hall 1857).

Charles Lyell had published the first of the many volumes and editions of his *Principles of Geology* in 1830, and his new ideas on the importance of modern sedimentary processes no doubt influenced Hitchcock's interpretations. Hitchcock (1841, p. 716–17) listed four types of sedimentary structures and their interpretations; these may be paraphrased as follows:

1. Parallel laminations: "quiet deposition on a level surface"
2. Waved laminations, usually ripple marks: "bottoms of rivers, lakes, and the ocean"
3. Oblique laminations [crossbedding]: "deposition on a steep shore"
4. Contorted laminations: result of "vertical and lateral pressure"

These elementary interpretations of primary sedimentary structures were a start toward understanding depositional environments. Unfortunately, although Hitchcock and his contemporaries were aware of these features, they did not fully understand or appreciate their significance and thus did not attempt to use them in their interpretations.

Hitchcock's later works (1858, 1865) acknowledged the growing evidence, based on relative dating of the fish and fern fossils, that the rocks of the

Hartford group (at least in part) were of Early Jurassic age. In this same respect however, he cautioned that the American stratigraphic sequence might not be perfectly analogous to the European (having been burned personally in this respect by his early assignment of the sequence to the Old Red Sandstone). Nonetheless, he still held out hope that workable beds of coal would be found in the Hartford basin.

Hitchcock became increasingly aware of the importance of sedimentary structures in the interpretation of environments. He eventually recognized that, when raindrop impressions and fossil footprints occurred on the same bedding surface, it was an indication that the surface had been subaerial rather than subaqueous; he then supported this idea with correct interpretations of desiccation cracks in the same mudstones (figure 1-6) (calling them "sun cracks"). These, in connection with the recognition of tree trunks in some of the sedimentary rocks, led him to the variable conclusion that the environment had been one of an "estuary, or lake, or river," yet he still preferred the conclusion that it had been the environment of an estuary "almost wholly cut off from the ocean" (Hitchcock 1858, p. 172).

Hitchcock's later works also played havoc with the taxonomic nomencla-

Figure 1-6. Fossil desiccation cracks on bedding planes of rocks from the Hartford group. Such sedimentary structures were considered to be primary evidence for a shoreline paleoenvironment, as muddy shorelines were then the only area where such cracks were known to be forming in modern environments. (Photo by the author)

ture, as he freely changed the names he had previously given to his trace fossils. As noted sorrowfully by Lull (1915), the rules of precedence in nomenclature published in 1844 had not gained general acceptance in Hitchcock's time.

Lyell: The Bay of Fundy Analogue

The estuarine theory, obliquely derived from the Wernerian marine/subaqueous ideas, became increasingly entrenched. This was in large measure due to American geologists' perceptions and applications of specific writings of Charles Lyell. Lyell visited America several times during the 1840s and chronicled many of his geological observations in a folksy travelogue. He attended scientific meetings, visited the Bay of Fundy estuary, and was shown several of the rift basins in North America by Hitchcock, Silliman, and Percival. He drew analogies between the ancient Newark Supergroup deposits and the modern Bay of Fundy, principally regarding the specific microenvironments that produced an assemblage of mud cracks, ripple marks, bird tracks, and raindrop impressions in the modern, intertidal zones. Lyell used his new principles of sedimentary processes in the interpretation of the rocks of the Hartford group, but he was constrained by concepts of marine-influenced sedimentation; ideas of entirely nonmarine sedimentation were still half a century away.

Such was the stature of Lyell, that the Bay of Fundy, complete with its 60-foot tidal range, quickly became the commonly accepted modern analogue for the ancient environments of each rift basin along eastern North America. Lyell, however, had only inferred that the muddy shoreline, with its alternate wet and dry (tidal) cycles, was an optimum environment for the creation and preservation of bird tracks, mud cracks, and raindrop impressions. Lyell's observations of similar processes and assemblages on the tidal shores of islands off Georgia were quickly forgotten.

Lyell also became convinced of the authenticity of fossil raindrop impressions during his visits to America, after seeing them in modern sediments and in the ancient rift-basin fills. He used the similarity in size of these impressions, modern and ancient, to argue against one contemporaneous theory that proposed a different density of the atmosphere in the past (Lyell 1851).

Rogers: The Theory of Primary Depositional Dip

One geologist who did not ascribe to either the estuarine or uniformitarian theories was Henry D. Rogers, the state geologist of New Jersey. He (and his brother, William B. Rogers, who worked in Virginia) had retained some of

the ideas of catastrophism well into the era when Lyellian–Huttonian uniformitarianism was gaining acceptance. Although Rogers worked south of
Connecticut, the rocks he studied were undoubtedly analogous to those of
the Hartford basin, and he did not hesitate to extend his interpretations
northward. This might have driven the more conventional geologists to
verbal mayhem, since the ideas he advanced were rather amazing. However,
the science was still at a stage where many theories could be advanced and
few proved or disproved, and most geologists remained reluctant to cast stones.

Rogers attempted to explain the present structural dip of the rocks as
primary depositional dip, the result of a large ("noble") river that had swept
along the coastal plain from south to north through the different basins,
emptying into the sea near New York City. Presumably a second river was to
have flowed south through the Hartford basin. His evidence for this river
and its effect on the sediments (1840, p. 167–69) was initially reasonable
enough, but his logic became progressively more strained, leading the
present-day geologist to suspect that Rogers had not visited many of the rift
basins outside New Jersey. His arguments also indicate the rudimentary state
of knowledge of fluvial processes that was prevalent at the time, although
Merrill (1906, p. 495) suggests that Rogers was in fact "wholly deficient in
theory" despite being "unequalled for the power of observation."

Rogers's evidence is examined here because it was offered in support of
such an unconventional theory and because, by inference, he applied it to
the Hartford group. Silliman's son, Benjamin Sillliman, Jr. (1844), in fact
made a preliminary attempt to apply Rogers's theory of inclined deposition
to the Hartford group, but his full report seems never to have been published.

Rogers (1840, p. 167–72) offered the following observations as evidence
that a fluvial system had deposited the rift-basin strata in an originally
inclined attitude:

1. In the ignorance of faulting and internal basin structure, Rogers calcu-
    lated that exposures of metamorphic basement rock within the basin
    demanded a thin covering of sedimentary basin fill, whereas geometric
    projection across the basin of the dip as a structural feature would
    "imply a thickness for the deposit so enormous as to be beyond all
    precedent. . . ." (p. 169).

2. Dip of the strata was constant, implying constant fluvial currents rather
    than fluctuating tidal currents. Local variations were seen as local
    eddies. Dips perpendicular to the axis of the basin were the result of
    "lateral influx" of the rivers and the oblique "across and down the
    channel" patterns of flow (p. 169–70). In this, Rogers confused dip of
    the strata with current-formed crossbedding.

3. The basins—and thus the fluvial deposits—were thought to become smaller southward toward the smaller headwaters of the supposed fluvial system. In addition, the regional northward decrease in elevation of the basins was equivalent to modern fluvial gradients, and therefore of similar origin.

4. Rogers also presented two rather opaque arguments that seem to suggest (a) that the "gradient" of the basins did not undergo the later structural uplift that affected the Tertiary deposits, and therefore could not have been a product of structural elevation, and (b) that, if the layers had been originally flat-lying, the conglomerates should always be at the base of the sequence, whereas erosional windows revealed shales below the conglomerates, arguing for inclined deposition. Neither of these arguments was clearly presented, and certainly neither would stand up under application of currently accepted principles of cumulative structure, facies relationships, or original horizontality of deposition.

At the 1842 meeting of the American Association of Geologists and Naturalists, Rogers further explained his ideas. Under questioning, he explained how evidence for the theory of inclined deposition could be found in the apparent lack of faulting of a type that would allow postdepositional tilting; he also noted the occurrence of some vertebrate footprints that seemed to indicate primary slopes by their sideways skidding motion, and he suggested that strong tidal currents associated with the alternative hypothesis (estuaries) would have disrupted any initially horizontal bedding.

When Silliman, in debate (Association of American Geologists and Naturalists 1843, p. 64), suggested that the structural tilt could have been related to "outbursts of trap" (unspecified mechanism), Rogers pointed out that the rocks of the Virginia basins were also tilted, yet there were no associated extensive flows. Lyell (p. 64) argued that the dip was too steep (often exceeding 20 degrees) for primary deposition, and that it was usually "transverse to the course of the ancient estuary." Rogers countered that the sediment had entered the basins from the sides, and that the shoreline had prograded into the basin during syndepositional uplift. This mechanism is also ambiguous, as read today.

According to Sanders (1974), much of the basis for this type of reasoning was a product of the inability of many nineteenth-century geologists to conceptualize or to accept the sheer magnitude of events possible in geology. This was probably related to their ignorance of the vast scale of geological time. If the dip of the strata in these basins was of primary sedimentary origin, then "inconceivable" thousands or tens of thousands of feet of strata need not have been postulated by the linear projection of the dips into the subsurface. This point of view now seems odd coming from the Rogers brothers, who were instrumental in deciphering the large-scale structure of the Appalachians, but such seeming inconsistency may only be the product of modern hindsight.

Rogers (1840, p. 171) also had attempted to explain the occurrence of conglomerates along the basin margins as the product of

> a violent agitation of the whole belt of country [an agitation that was, interestingly enough, supposed to have been caused by an "outbursting of trap"], and the vertical rising of the bed of the red shale valley to a higher level, which would necessarily set into motion the entire body of its waters. These, rushing impetuously along the shattered strata of the base of the hills confining the current to the northwest, would quickly roll their fragments into that confused mass of coarse, heterogeneous pebbles which we see, and strew them in the detached beds of con- glomerate which they now form.

These "stupendous changes" were to have taken place over a brief interval of time, but (as noted earlier) they were evidently not presumed to have been associated with any faulting or permanent structural deformation—an inter- esting geological non sequitur.

Rogers also opposed Lyell's uniformitarian ideas about rates of tectonic movement. Lyell (1843) had argued for slow rates of continued subsidence over extensive periods of time to produce large magnitudes of subsidence in the basins. He used, among other evidence, observations of the thinness of layers of modern sediments containing bird tracks in contrast to the thick accumulation of such thin beds of strata containing tracks that Hitchcock had shown him in Massachusetts. Rogers, in another example of particularly opaque scientific writing (1856), argued against synsedimentary subsidence by suggesting that, since ripple marks were known to form in deep water, rippled sediments that did not contain "such unequivocal shore marks as those of footprints" (p. 315) could well be of deep-sea origin. Therefore, a deep basin could have existed a priori and could have filled slowly thereafter without contemporaneous (slow) subsidence.

While Rogers's principal theory of primary depositional dip was correct in proposing (for the first time) a nonmarine model, the ideas were inconsis- tent and incorrect in their own right. They received little credence from most geologists working in the Hartford basin, who often felt compelled to defend their views and to offer evidence opposed to the idea of primary depositional dip.

Percival: The Geology Map of Connecticut

Two reports on the geology of Connecticut were written in the early 1800s. The first, published in 1837 by Charles U. Shepard, M.D., was a short but detailed report on the economic geology of the state; as such, it dealt little with the poorly mineralized rocks of the Hartford group.

The second report was published in 1842 by James G. Percival. It was to have been a companion report to Shepard's (they had both been commissioned by the state), but it did not turn out that way. Percival was (or felt himself to be) underfunded, and he petulantly let the world know about it in print. In fact, he had missed so many deadlines that his funding was finally terminated by the state legislature. His final report resembled other massive tomes of the day, consisting of 500 pages of dense prose and one map (figure 1-7). But the science of his report fares poorly when compared to most contemporaneous geological reports. This is especially true of Percival's report in comparison to Hitchcock's *Final Report on the Geology of Massachusetts*, which had been published the previous year.

Percival was a collector rather than an interpreter. His penchant for collecting was alluded to by Dana (1891*b*), who noted that, although Percival died in virtual poverty, he had collected a library worth $20,000. Certainly he must have been knowledgeable in order to collect the quantity of precise geological observations he needed to produce his accurate geological map of the state, but he seldom allowed conclusions or interpretations to enter his writings. He had, as noted by Dana (1891*b*, p. 439), a "memory which retained all the facts that ever entered it," but as Mark Twain wrote (1896, p. 67) of a similar man, "Such a memory as that is a great misfortune. To it, all occurrences are of the same size. Its possessor cannot distinguish an interesting circumstance from an uninteresting one."

Percival's report is a collection of minute geological descriptions correlated with detailed geographic locations; as such, it was an excellent foundation for a geologic map, but a poor one for understanding the geology described. Most of his contemporaries seem to have realized this and were either silent concerning his report or as time went on made such references as Davis's (1886, p. 352) to "the mazes of Percival's text." Later authors, apparently despairing of wading through the text, often assumed there were nuggets to be gleaned somewhere but did not bother to try to find them, and referred to the report in glowing terms. In some respects, we must be thankful that the state legislature voted him funds sufficient only to write what he termed a "hasty report" of a mere 500 pages, rather than the full-length report he had envisioned.

Percival did make a major contribution, however, in his accurate geologic map of the state. This was a marked improvement over Hitchcock's 1823

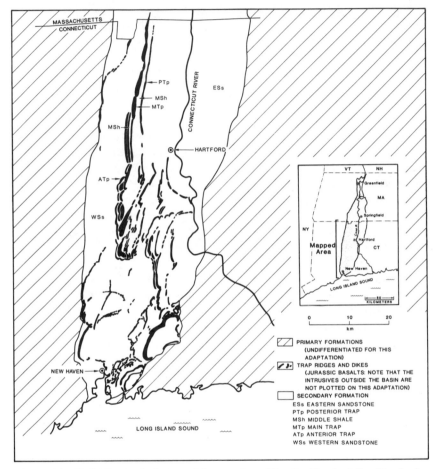

Figure 1-7. 1842: Geological map of the southern Hartford basin (after Percival 1842). Percival's mapping was exceptionally accurate for the time and served as the basic geological map of the Hartford basin in central Connecticut for over half a century. Note that Percival did not distinguish between intrusive and extrusive basalts. Both are reproduced here, whereas in later maps presented in this book the intrusives are omitted as complexities that (because they had been distinguished by contemporaneous geologists) would be easily factored out of the sedimentary history of the basin.

version and provided a solid basis for future interpretations. Percival divided and gave geographically descriptive names to the different parts of the basin fill, based on their positions with respect to the basalt flows and on their geographic locations with respect to New Haven. Thus, in ascending stratigraphic order (although it is not clear that Percival had grasped the concept of stratigraphic order), Percival's map showed the "Western Sandstone," "Anterior Trap," "Middle Shale and Main Trap," "Posterior Trap," and "Eastern Sandstone," and this terminology remained in use for some eighty years.

Percival did note that the darker rocks and the shaley intervals seemed to be closely associated with the basalt flows, but he offered nothing more than the observation. He also provided the insight, spread out in several different passages, that the plant fossils in the rocks were unlike the "ferns" known from the Carboniferous coal measures of Europe, and suggested a lithologic correlation with the New Red Sandstone (Permian–Triassic), or perhaps the Bunter sandstone (Lower Triassic) of Germany.

Descriptive and noncommittal as he was, the accuracy of Percival's 1842 map was years ahead of Hitchcock's maps of Connecticut (1823) and Massachusetts (1844). Hitchcock had not, as Percival claimed to have done, repeatedly traversed his state east to west in parallel tracks only two miles apart. Hitchcock was still speculating in 1858 about whether the similar shale sequences above and below the Holyoke Basalt might have been connected in an arch over the basalt as a single formation—apparently never understanding the stratigraphic significance of the uniform easterly structural dip of the two shale sequences and the basalt. Perhaps fortuitously, Percival's map was based on a stratigraphic division of the rocks, whereas Hitchcock's were based on lithologic differences in a sequence in which lithologies are distributed unevenly and are repeated at several stratigraphic intervals.

Percival apparently knew more than he was prepared to write about. A short passage of Lyell's (1845) credits Percival with demonstrating to Lyell personally the contact metamorphism both above and below the dikes around New Haven, proving their intrusive nature. This same proof is presented vaguely at best in Percival's own report. Percival's ability to interpret geology was also mentioned briefly in a review of his report in the *American Journal of Science* (1843) and even earlier by Hitchcock (1823); but whereas Hitchcock and his ideas grew with the science of geology, Percival seems to have been left bewildered when the rocks did not match the preconceived ideas of the day, and he took refuge in a purely descriptive mode of writing.

Redfield: The Age of the Deposits

The other workers who were actively involved in the interpretation of the Hartford group prior to 1850 were John H. Redfield and his son William C. Redfield. Beginning in 1836, the Redfields began publishing a series of paleontological descriptions of the fossil fishes of many of the rift basins along the east coast. They quickly reached the conclusion that the fish correlated with post-Permian rocks of Europe, and argued against correlations with either New Red Sandstone or Old Red Sandstone. By 1856, they had decided that the fish could be ascribed to the latest Triassic and/or Early Jurassic (Liassic) of the existing European stratigraphy.

The Redfields' work later fell into disrepute among geologists who believed that the plant and other vertebrate fossils were more compatible with a Late Triassic age and that the fish could not be so specifically age-diagnostic. By the end of the 1800s, the deposits were almost universally ascribed solely to the Late Triassic. Recent palynological and radiometric age-dating, as well as new work on the fish themselves (discussed in Part II), have revived and supported the Redfields' earlier interpretations.

2 / Late 1800s

After Percival's and Hitchcock's reports on the geology of Connecticut and Massachusetts, no major changes or advances occurred in the description and interpretation of the Hartford group for some thirty-five years. Dana used the rocks of the basin to illustrate many geological concepts in his influential *Manual of Geology*, giving his own twists to the interpretations, but he did little original work in the basin, and his ideas were rarely the subject of discussion. These rocks were old hat and generally of far less interest than the new, fascinating, and exceptionally well-exposed outcrops being discovered and described in the course of exploration of the Rocky Mountain areas. Besides, Dana was a geologist of some stature, and it could be dangerous to get involved in an argument with him. Having married Silliman's daughter and taken the Silliman chair at Yale, Dana had also taken over much of the editorship of the *American Journal of Science;* and this, combined with his recognized geological ability, put him in a very influential position despite his ill health in later years.

Dana: Equivocal Interpretations and the *Manual of Geology*

First published in 1863, Dana's *Manual of Geology* reinforced the estuarine theory in several ways. Dana's conception of the environmental significance of primary sedimentary structures led him to support a beach, tidal, or eolian type of environment for any kind of crossbedded sandstone. Most of the smaller sedimentary structures that he recognized (ripple marks, "wave marks," "rill marks") were also associated in his mind with a shore-line environment.

Henry C. Sorby (1859) had just published a paper on the environmental significance of such sedimentary structures, and Dana (1863, p. 425) proceeded to use such structures in the Hartford basin in conjunction with the fossil evidence to infer a generalized depositional environment of "tides or fresh-ets" along a basin "partly of estuary and partly of lacustrine" origin. Dana dismissed the possibility of "seashore" environments because of the absence of marine fossils. Thus, his conclusions remained ambiguous and contradictory. He went so far as to note that the absence of seashore deposits in these basins adjacent to the Atlantic was strange, and postulated that an extension of the land seaward might have accounted for it. Having no idea that Africa had been welded onto North America at the time, he speculated that a drop in sea level of 500 feet (170 meters) would have extended dry land some 140 kilometers seaward, leaving the ocean "imperfect access, if any" (p. 439) to an estuary in the position of the Hartford basin.

Dana, like the other geologists of his era, was struggling to force nonmarine data into a model of marine-influenced environments. Despite his stature as a geologist, Dana was doing this less elegantly (or at least in a less consistent manner) than others were. Glacial deposits were then in vogue, having been recognized several decades earlier, and Dana even invoked glacial conditions and violent floods in his later publications (1879, 1883) in a flailing attempt to account for the eastern border-fault conglomerates. Once in a while he let slip a thought that the deposits "correspond well, in all parts, to those of fluvial and estuary origin" (1883, p. 384), but he almost always returned to the primarily estuarine model for lack of anything better. In the fourth and last edition of the *Manual of Geology* (1896), Dana briefly considered fluvial, estuarine, and glacial environments of deposition, but made even fewer attempts to document interpretations than he had in the 1863 edition. A separate discussion of modern alluvial "cones" noted that "such fresh-water accumulations have thus far been found only among recent formations" (1896, p. 99) and made no attempt to relate them to any ancient strata.

In another area, Dana (1873) gave the name *geosynclinal* (later to evolve into *geosyncline*) to a concept that James Hall had developed fourteen years earlier. This concept envisioned basin formation, its filling with sediment, and the final formation of mountains in the same area, as a universal sequence of events in mountain-chain development. The geosyncline concept and its variations strongly influenced geological thinking regarding basins for three quarters of a century. Many aspects of it are still valid, although the underlying mechanisms are better understood in light of plate tectonics.

In this 1873 paper, Dana suggested that the Hartford basin was an example of a truncated geosynclinal cycle: the red coloration of the sediments was seen as evidence for pervasive heat during early metamorphic stages of the cycle. The full metamorphic stages of burial—and the ensuing formation of mountains as called for in the geosyncline theory—had supposedly never been attained in the basin, and therefore it was an incomplete cycle.*

Russell: Broad Estuaries and Red Color

Broad Estuaries

Israel C. Russell was an outsider to the rift basins of the east coast, working at different times as a photographer in New Zealand, a geologist with G. K. Gilbert (working on Pleistocene lakes of the west for the U.S. Geological Survey), a professor at Columbia University, and finally a professor at the University of Michigan. Because his ideas were unorthodox and because of his "foreign" status, he probably received more than his share of criticism. Yet he persevered and did a better job than Dana had of fitting square-peg nonmarine data into the round hole of an estuarine depositional model. In effect, Russell was wrong for the right reasons. Russell probably made the best use of the sedimentological data and models available in his day, but he also reached some of the most inaccurate conclusions.

Russell took the broad-brush approach, studying the Hartford group only as an element of the system of deposits and basins along the east coast of North America. His first major publication on the subject appeared in 1878 in the *Annals of the New York Academy of Science*. Russell disagreed with Dana on several important points, and he rarely published in the *American Journal of Science*, which was under Dana's control.

As did others before him, Russell (1878, p. 225) started with the premise that "it is well known that ripple-marks on rocks indicate that they were deposited in shallow water, and that shrinkage-cracks, raindrop impressions, and the footprints of animals are formed where broad areas of muddy shore are exposed at low tide." The logic of the first half of this sentence stands up to today's scrutiny, but then Russell, like others, confused correct first-order interpretations of microenvironments with second-order interpretations of environmental systems. The geologists of his day considered these first-order sedimentary structures to be restricted to microenvironments in a single, second-order (marine) system. Russell then compounded the error by concluding (p. 225) "that the rocks bearing these inscriptions were deposited

---

*Kay, the prime proponent of the theory of geosynclines seventy-five years later, would take Dana's concept one step further (1944, 1951). By the 1940s, the fault-bounded nature of the Hartford group was known; but because it was one of Dana's examples, Kay included it as a type of geosyncline and coined the term *taphrogeosyncline* for this class of faulted basin. This was one of the first among the many types of geosynclines recognized and named by Kay as the term became increasingly inflated and unwieldy.

in a body of water subject to high tides," and by describing the Bay of Fundy as the modern analogue to the Hartford basin.

The Bay of Fundy (figure 2-1) was a seductive setting to the early geologists. It offered them a preliminary explanation for the ancient red sedimentary rocks, in that the red rocks being eroded by waves all around the bay were providing red detritus to the adjacent depositional area (Klein 1963). As had been shown by Lyell, a geologist of some stature, it contained mud flats wherein subaerial bird tracks and raindrop impressions could be formed, dried out, mud-cracked, and then quickly buried and preserved through the action of high tides. High tides also offered a mechanism by which the evidence of subaerial exposure could be readily juxtaposed with supposedly marine sediments. Until more knowledge was available and rigorous geological proofs could be used, Ralph Waldo Emerson's principle that "people only see what they are prepared to see" applied to the interpretations of the Newark strata and the Bay of Fundy.

It is difficult for us to appreciate the limited nature of the information available to these early geologists. Most of them did not realize that the ancient rift basins were not near an obvious source of red sediments, as is the Bay of Fundy. (When that fact did become obvious, there followed decades of dominance by the theory of primary red coloration of the sediments that

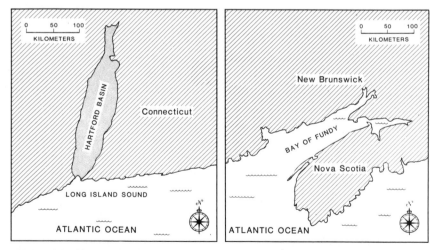

Figure 2-1. Comparison of the superficial features of the modern Bay of Fundy and the Hartford basin. The modern estuary was taken to be a depositional environment analogous to that of the Hartford group. Apparent similarities included access to the open ocean, deposition of red sediments, and a similar suite of sedimentary structures.

form redbeds.) Nor did the early geologists know that the sedimentary structures were not-tidal (especially not macrotidal) in origin.

The cause underlying the origin of "strata, or the so-called division of sedimentary matter" (Wells 1852, p. 297) was still debated. Sorby's work on sedimentary structures was not published in a collected and widely accessible journal until 1908, and even so his principles were not generally applied until decades later. Such concepts of the environmental significance of individual or suites of sedimentary structures are the foundations on which we presently build our sedimentological interpretations, yet they are so familiar that few present-day geologists appreciate the laborious efforts that went into their formulation and acceptance.

For Russell, the high tides of the Bay of Fundy and the associated strong currents also explained the border-fault conglomerates of the Hartford group. He took the conglomeratic sediment and conceptually spread it out along an estuarine shoreline, identifying the high energy of exceptional tides as the agent capable of transporting the coarse-grained sediments.

Previously, these had been lamely explained as glacial deposits (Dana) or had been ignored. The faulted nature of the eastern margin of the basin and its role in producing conglomerates had not yet been recognized. Russell, however, was familiar with the pebbly (glacial remnant) beaches of the shores of the North Atlantic and the Great Lakes. He also recognized, from modern examples, that steep "impetuous" streams were capable of carrying coarse, pebbly sediment down from adjacent highlands to be deposited onto adjacent flatlands. From this he correctly concluded that streams had transported the ancient cobbles down from basin-margin highlands, but he envisioned them as having been supplied to a shoreline environment. There was as yet no model of subaerial alluvial-fan deposits in the literature, however, and although Russell himself (1885) had described modern arid fans from the American west, he continued to labor within the confines of the estuarine model.

Russell then asked why the sediments were not as coarsely conglomeratic at the western edge of the Hartford basin or at the eastern edge of the Newark basin, and concluded that these basins represented opposing halves of his high-tidal estuarine model (Russell 1878, 1880). Other data then could be interpreted easily according to this newly-formed, broad-terrane hypothesis, in which the Connecticut and New Jersey formations were erosional remnants at the opposite margins of a much broader estuary.

The supporting data included the similar stratigraphic sequences in the two remnants, complete with several layers of basalt in each. The opposing dips of the two areas (figure 2-2), were interpreted as being the result of regional postdepositional uplift that occurred between them. The clinching piece of evidence was the existence of the Pomperaug outlier midway

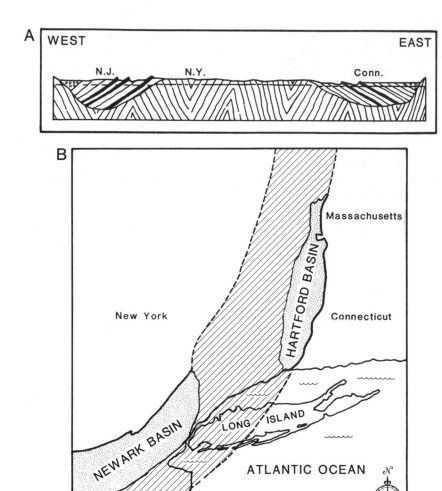

Figure 2-2. 1878: (A) Russell's initial evidence for the broad-terrane hypothesis: opposing dips and similar stratigraphic sequences. (After Russell 1878, Figure 1, p. 230) (B) Sanders's later concept of the configuration of the greater Newark/Hartford broad-terrane basin. Note that Sanders subscribed to the broad-terrane hypothesis, not to the estuarine depositional environment. (After Sanders 1963, Figure 2, p. 503. Reproduced with the permission of the *American Journal of Science*)

between the two main remnants, proving that formerly the Triassic deposits had indeed been more extensive.

Perhaps the most glaring error in Russell's model was the absence of shoreline conglomerates in his modern analogue, the Bay of Fundy. Other geologists, however, undertook to attack his reasoning on other grounds because their concepts of depositional models were not sufficiently refined to be used as a basis of argument. Although the broad-terrane hypothesis is not generally accepted at present, Dana's objection that Russell's model relied on excessive magnitudes of subsidence and uplift in the intervening areas between the basins is no longer valid. Vast amounts of subsidence and uplift are currently recognized in many basins, but the sheer magnitude of tectonic events was beyond the bounds of experience of the early geologists.

Red Color

Russell (1889a) broadened his scope of interest with a comprehensive review and hypothesis on the origin of the red coloration of the Newark Supergroup.

This problem had been mentioned briefly by earlier writers. Hitchcock (1841, p. 526) spent a paragraph suggesting that the red color "undoubtedly . . . proceeds from the red oxide of iron, which, in some way, has diffused through the mass." The diffusion agency Hitchcock preferred was that of ill-defined syndepositional heating "while yet at the bottom of the ocean." Dana (1863, p. 432) briefly addressed the problem, writing that "the red color of the sandstone—a consequence of oxydation [sic] of magnetic-iron grains present in it—appears to have had its origin [during a period of volcanism]." He later speculated that the iron itself may have had a volcanic source.

Russell (1889a), after reviewing many of these early theories, put together one of the better-reasoned arguments of the era. According to Russell, the red color was the result of three factors: the dissolution of iron silicate minerals in the source area, the oxidation of the liberated iron within the thick red soils in the same area, and the transportation and deposition of the iron as an oxide. Finally, he proposed that this chemical activity indicated a warm, humid climate at the time of deposition, using the climate and lateritic soils of the southeastern United States as a modern analogue.

Again, Russell was wrong on all counts according to present evidence, in part because his interpretations were hampered by a lack of knowledge of postdepositional diagenetic processes. Although some investigators had earlier suggested the derivation of red coloring material from the in situ decomposition of iron silicate minerals, most felt that the resulting volumes of released iron were too small to provide the observed redness of the rock, and, more importantly, that some subsequent agency of heat was necessary to oxidize the released iron.

Postulated heat sources were either too localized (volcanic intrusions) or too widespread (metamorphism) to account for both the observed redness of entire formations and the bed-by-bed redness of variegated formations respectively. Workers such as Russell also had to account for high volumes of detrital iron oxide, since the supposed modern analogue, the Bay of Fundy, could be seen to be depositing locally derived red sediment. Moreover, the iron had to have been oxidized prior to transport and deposition, given the numerous occurrences of drab formations that had high iron content; thus, the iron must have been oxidized in the source area. This idea was supported by microscopic observations of discontinuous red coatings on grains (interpreted as evidence for partial abrasion during transport of red coatings acquired in the source area). Additionally, the observation of grains supported by "ocherous matter" strengthened these geologists' impression that the red matter was of syndepositional origin, inasmuch as the only alternative under considera-tion was postdepositional infiltration of material.

From this, Russell constructed the first well-reasoned, well-supported (yet wrong) climatic model for the Newark Supergroup's depositional setting: that of a warm, moist climate, wherein the source rocks were rapidly decomposed and concurrently oxidized in situ. This model contradicted the European tradition of interpreting the Triassic as an arid era, and as a result it did not gain much of a foothold in the European scientific community, but Russell's ideas would profoundly influence American thought on the origin of redbeds for almost three quarters of a century.

Other Ideas

In 1878, Russell briefly stated a concept that had probably been recog-nized implicitly by many geologists, but that had been poorly understood, seldom used, and often openly denied—especially in America. He had a glimmer of the concept of facies equivalents (figure 2-3) and wrote that the conglomerates in the Hartford basin had been "formed contemporaneously with the beds of shale and sandstone" (p. 236). It was an idea Russell did not elaborate on, and its significance was probably never appreciated, although Dana (1896, p. 744) had also grasped the concept when he stated that, in the Triassic basins, "conglomerate-beds, sand-beds, and mud beds may have been forming simultaneously at the same horizon in different portions of the area." Until nonmarine depositional models and a better understanding of sedimentary structures existed, however, such ideas were not used in any significant environmental or paleogeographic reconstructions of the Hart-

"SHORE BLUFFS"

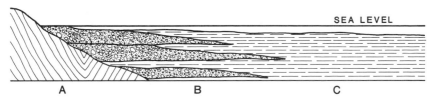

Figure 2-3. 1878: The concept of facies-equivalent lithologies was used by Americans in the Hartford basin as early as 1878, but the potential of this mental tool was not fulfilled until much later. Note the lack of a faulted basin margin. (A) "Shore bluffs"; (B) conglomerates "brought down from the highlands"; (C) "the sandstones and shales which are the ordinary shore and offshore deposits." (After Russell 1878, Figure 2, p. 235)

ford group. The concepts of facies assemblages and facies-equivalent lithologies were being developed and used at this time in continental Europe, but they would not be profitably applied to the Newark rocks for some time to come.

Russell's final contribution to the study of these deposits was to reinstate permanently William Redfield's 1856 *Newark Group* terminology. He suggested this in 1889 (1889*b*) and subsequently (1891, 1892, 1895) had to defend his proposal, as others made an issue of names and priorities. The terminology has since been accepted, although it has been expanded to the term *Newark Supergroup,* to include all of the rift-basin deposits along the Atlantic coast of North America (Van Houten 1977; Olsen 1978).

Davis: Structure and Nonmarine Deposition

William M. Davis was one of the last influential geologists to address the problems of the Hartford basin in the nineteenth century. His approach made use of an interesting combination of structure and geomorphology, and he derived much support for his famous theory of peneplains and cyclic geomorphology from his study of the Hartford group.

Beginning in 1882, Davis began to publish his work on the unraveling of the postdepositional structure of the Hartford basin. His main contribution was the documentation of the fault repetition of the stratigraphic sequence, based on the correlation of similar stratigraphic sequences between the different fault blocks.

Davis, a professor at Harvard, also carried on a low-key, running feud with Dana at Yale regarding the intrusive versus extrusive nature of most of the basalts. Dana's (1891*a*) ideas in favor of intrusion were considerably outdated; because he had (correctly) documented the intrusive nature of basalts in the New Haven area, he never felt the need to check the other basalts, which are predominantly extrusive. Additionally, Davis speculated on the origin of the Hartford basin as a compressionally folded downwarp, closely related to (but slightly younger than) the formation of the Appalachian Mountains. This latter idea coincided with Dana's concepts but was destined to be superseded rather quickly.

Structure

Davis's recognition of postdepositional fault blocks was facilitated greatly by Percival's 1842 map and report. Although Percival, consonant with his reluctance to offer interpretations, had not mapped basalt ridges any farther than he could actually see them cropping out at the surface, he had postulated only three basalt units. Davis extended the basalts to inferred fault lines, and began to map faulted offsets (figure 2-4).

In some respects, Davis seems to have made the problem and its interpretations more complex than necessary. Although no one had suggested it, he proposed (1888, p. 463) that "the numerous interruptions in the ridges formed by these trap outcrops might be regarded as indicating so many independent intrusions," and he then set about proving otherwise in detail. His prime evidence for correlation lay in the stratigraphic sequence, where it could be shown that a "limestone" bed always occurred in the lower part of a shale unit that separated two basalt flows, and where he and Loper (1891) demonstrated the extent and former continuity, offset by faults, of two distinctive beds containing fossil fish.

It was important to Davis's interpretation that the majority of the basalts be subaerial flows rather than intrusive sills, in order for him to use the flows as sequential parts of the stratigraphic sequence and as structural marker beds. He recognized that he could not have used erratically placed sills in this way; so he went to some trouble to document the eruptive nature of the basalts.

In definitively sorting out the postdepositional structure of the basin, Davis made a valuable contribution to future sedimentological and stratigraphic interpretations by making it possible to subtract these complications from the stratigraphic sequence. He also opened the way for extensive use of the concept of facies-equivalent lithologies in his definitive proof that most of the strata-bounding basalts were time-line subaerial flows. Because the basalts were important to his theories, he studied them closely and was the first geologist (1896) to recognize the distinctive *aa* type of flow in the ancient basalts of the Hartford group. The identification of pillow basalts would not occur for several more years.

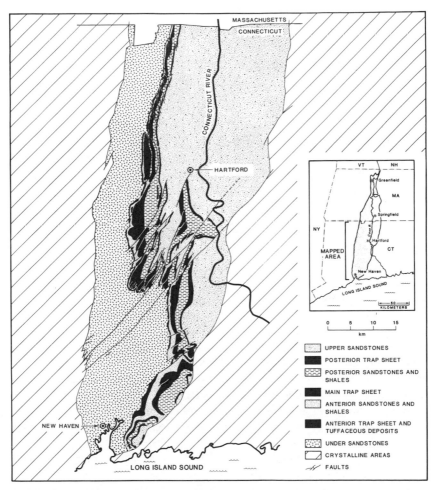

Figure 2-4. 1898: Davis's map of the Connecticut half of the Hartford basin (after Davis 1898, Plate 19). Davis's map showed the same basic patterns shown by Percival's 1842 map—albeit with considerable refinements in detail—but Davis did make two important advances. First, he mapped faults, in order to account for offsets in the basalt ridges; and second, he distinguished between intrusive basalts (not shown in this adaptation) and extrusive basalts. Davis's descriptive stratigraphic nomenclature was similar to Percival's, except that he used the stratigraphically significant terms *Upper* and *Lower* in place of Percival's geographically descriptive terms *Eastern* and *Western*.

Nonmarine Environments

Davis's ideas on depositional environments provided a transition from nineteenth-century conceptions to those of the early twentieth century. At first, the interpretation of environments did not interest him, and he followed conventional thought rather closely in his early, general references to "sub-aqueous" deposits. In his final report (1898), however, he devoted several pages to mulling over the different possibilities. He finally decided that the weight of evidence—consisting primarily of the absence of marine fossils and the potentialities of the newly recognized (though ill-defined) nonmarine depositional processes—suggested that the term *continental* "seems more applicable than any other to the Triassic deposits of Connecticut. It withdraws them from necessary association with a marine origin, for which there is no sufficient evidence, and at the same time, it avoids what is today an impossible task—that of assigning a particular origin to one or another member of the formation" (1898, p. 33).

This remarkable, though still lukewarm statement, constituted the first real break from the old estuarine model and was probably elicited from Davis by two factors. The first was the emerging concept of nonmarine sedimentation, promulgated in Europe by Penck and Walther. The second was the continued exposure of American geologists to the modern nonmarine deposits of the American west during government-funded geographic surveys. The latter directly influenced Davis, who had himself spent time in Montana in the early 1870s. Davis's work on the Hartford basin during the 1880s was also done under the direction of R. Pumpelly and G. K. Gilbert of the U.S. Geological Survey, both of whom were important figures in many of the expeditions to the territories of the western United States. Davis took the new ideas, and concluded that the conglomerates represented stream deposits, that the shales represented "quieter conditions," and that there may have been shallow lakes along the middle of the basin.

Although many geologists continued to cling to the estuarine theory, the first step had been taken, and the way was now opened for more realistic interpretations of the depositional environments of the Newark Supergroup.

Basin Structure

In another area, however, Davis's ideas did not withstand the test of time. Davis believed that the Hartford basin was originally an unfaulted compressional downwarp of the crust, and that postdepositional tectonics had caused all of the presently observed faults (figure 2-5). In his arguments, Davis confused synsedimentary and postdepositional structures (Davis 1898; Davis and Griswold

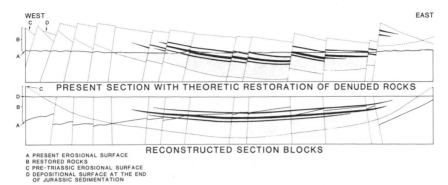

WEST   EAST

PRESENT SECTION WITH THEORETIC RESTORATION OF DENUDED ROCKS

RECONSTRUCTED SECTION BLOCKS

A PRESENT EROSIONAL SURFACE
B RESTORED ROCKS
C PRE-TRIASSIC EROSIONAL SURFACE
D DEPOSITIONAL SURFACE AT THE END
  OF JURASSIC SEDIMENTATION

Figure 2-5. 1898: Davis's concept of an originally downwarped basin that had been postdepositionally faulted (after Davis 1898, Plate 20). Davis confused important evidence for syndepositional faulting with the more common evidence for postdepositional faulting.

1894), suggesting that—because the oblique faults obviously could not have occurred during deposition—the basin margin could not have been faulted during deposition, either.

Davis hedged his bets (1898, p. 39) by stating that "it does not seem advisable at present to make final choice between these alternatives" (down-faulting versus downwarping); but all his arguments were pointed toward supporting the latter idea. It must be remembered that all of the faults—both border faults and oblique faults—were still inferred at this stage. Although the fault offsets in the basalt ridges are apparent on maps, no fault gouge or slickensides and (at the eastern border) no offset or sense of motion are exposed in natural outcrops.

Emerson: The Geological Map of Massachusetts and Marine Interpretations

The same year Davis's final report was issued (1898), B. K. Emerson published new maps (1898a, 1898b) of the central part of Massachusetts. Emerson, a professor at Amherst, had been inspired to take up geology by Hitchcock. Emerson became an eminent stratigrapher, but his concepts of depositional environments of the Hartford group were still caught up entirely in the estuarine model and the Bay of Fundy. He interpreted all of the different lithologies as synchronous facies, interpreting the conglomeratic Mount Toby Formation and the Sugarloaf Arkose as shoreward deposits, and interpreting the Longmeadow Sandstone and the Chicopee Shale as progressively offshore (and quieter) water deposits. Within limits, this was valid and a commendable use of facies concepts, but Emerson extrapolated these ideas to the entire stratigraphic sequence, regardless of structure or time lines.

It was not an arbitrary decision to do so, but one based on his previous work (1891), in which he had decided that the extreme lateral variability of the beds in these clastic lithologies precluded a standard layer-cake stratigraphic division of formations. Therefore, he had decided to attack the problem by "for the time disregarding dip and strike" in favor of trying "to map the area [by] making lithological distinctions" (Emerson 1891, p. 452). This approach contrasted markedly with Davis's primarily structural methods and produced a cumbersome concept of the stratigraphy—one that would bias his interpretations in all his subsequent work.

In essence, Emerson believed that the conglomerates exposed in the western part of the basin were a basin-margin facies synchronous with that of the eastern border conglomerates, despite the fact that they are stratigraphically separated by an isochronous basalt flow. In fact, Emerson felt that the consistency and importance of the structural dip, which would have shown his facies superimposed, was "overstated." He did have some precedent for this, in that Hitchcock had earlier made little use of the structural dips in Massachusetts and had speculated on the correlation of sandstones on the east and west sides of the basin, but it led to an ambiguous understanding of the Hartford group in Massachusetts.

Thus, while Percival and Davis had mapped the Connecticut formations on the basis of position with respect to the basalts—a system by which (fortuitously) each formation is roughly a time-rock unit—Emerson separated the Massachusetts formations on the basis of similar lithology (figure 2-6). As a result, time lines in this part of the basin crossed the same formation boundaries a number of times.

Emerson fit all of these data into a nebulous estuarine depositional environment. Influenced by Dana's *Manual of Geology,* he wrote of "shore ice" that was inferred to have created the coarse, angular border conglomerates. Direct evidence of glaciers was missing, therefore Emerson relegated them to the neighboring highlands. He noted evaporite pseudomorphs in the mudstone facies but did not yet try to interpret their environmental significance. Finally, because many of the footprints, mud cracks, and other subaerial indicators are found in the central-basin "deep water" deposits, Emerson was forced to postulate an elevated, sometimes emergent mud flat in the middle of his estuary.

Emerson also noted an apparent disparity between the mineralogy of the conglomerate clasts and the composition of the immediately adjacent highlands. In response, he postulated strong tidal shoreline currents to account

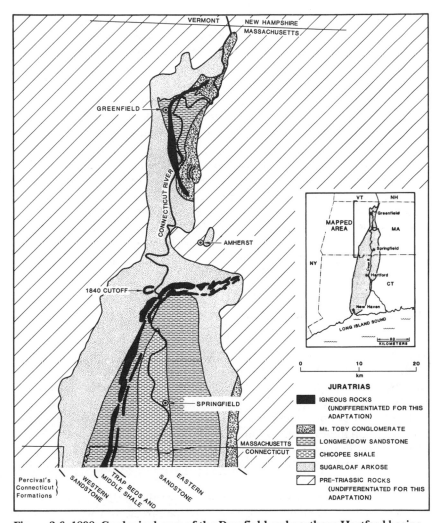

Figure 2-6. 1898: Geological map of the Deerfield and northern Hartford basins (after Emerson 1898). Emerson mapped the formation almost exclusively on the basis of lithology, ignoring time-line basalt-flow units and structural dip. Three different igneous lithologies are distinguished on the original map, but they are not differentiated here for lack of space. Note the noncorrelation of Emerson's formations with those to the south across the state line. Emerson mapped these rocks as "JURATRIAS" strata. Emerson's 1917 geological map of the same area was essentially unchanged from his 1898 map.

for what he perceived as a 30-kilometer lateral offset between clasts and their source areas. He even cited some of the first crossbedding observations (1898*b*, p. 375) to support his ideas of northward paleotransport, although more recent quantitative measurements in the area suggest that his observations may have been incorrect.

Emerson (1898*b*) apparently had some insights into the structure of the area, using the term *graben* in its first association with the Hartford basin. It is not clear, however, whether he inferred the faulting to have taken place syndepositionally or postdepositionally, as he alternately used the terms *syncline* and *graben* and never discussed any implications of the faulting.

In his later (1917) publication (discussed below), Emerson was to change some of his ideas on depositional environment, but not his basic map. As a result, a "borderline fault" existed in the geology between Connecticut and Massachusetts for many years, and it required some mental effort to correlate the geology between the states. Despite its shortcomings however, this type of map did portray the limits of some of the lacustrine units in the northern part of the basin.

Hobbs: The Pomperaug Outlier

Although he published in 1901, Hobbs did his work in (and his style of interpretation belongs to) the Russell and Davis era. Hobbs wrote an impressively long U.S. Geological Survey report on the Pomperaug outlier, a subbasin that is postdepositionally detached from the southern part of the Hartford basin (see figure I-3).

The report did not offer any new interpretations other than the suggestion, based on mineralogical affinities with possible source rocks, that much of the subbasin fill was derived from an area between the outlier and the main Hartford basin. This interpretation has not been upheld by subsequent work.

Hobbs also provided some basic data on stratigraphic thicknesses and lithology that would be used in later interpretations, as well as a descriptive section on petrology and a version of Davis's fault-block, postdepositional deformation that is amazing in its complexity.

Regarding depositional environment, Hobbs thoroughly reviewed all of the earlier ideas, including Davis's nonmarine scenario. He was noncommittal as to which side he favored, except in a short passage (1901, p. 42–43) where he noted an intraformational conglomerate of mudstone rip-up clasts. These, he suggested, were best explained by synsedimentary processes in the course of which "certain portions, partially consolidated, were locally elevated above the general surface" of the water. He was still a proponent of the estuarine model, however, when he published a paper (1902) in support of Russell's broad-terrane hypothesis. He seems eventually to have changed his ideas, as Barrell (1906) cited a talk given in 1905 wherein Hobbs advocated nonmarine depositional environments for these deposits.

3 / Early 1900s

Beginning in the early 1900s, the style of performing and reporting geological research changed. Geologists no longer devoted years on end to a single project, and the number of papers and the number of pages per paper by an author on a given subject decreased. Concurrently, individual geologists were developing a wider variety of interests, and more geologists were being trained, so the number of different authors who published on the Hartford basin increased.

Along with the new ideas of nonmarine deposition that were developed and brought back from the American southwest, new data became available as road cuts were made and aqueducts were tunneled through the rock. These provided fresh exposures of rock, sediment/basalt contacts, and faults, confirming many previous inferences but altering others.

The magnitude of absolute time was finally established with the emergence of radioactive dating, and geologists could begin to talk cogently about rates of deposition and erosion. Although developed earlier, sedimentary petrology was finally put to good use in the 1920s, and sedimentary structures began to be used in a semirigorous way slightly later.

These changes were reflected in the interpretations that began to be made in the Hartford basin. Yale continued to be the center of much of the study, and—now that the western territories were becoming states and developing their own geological surveys—interest gradually refocused on the home ground around the eastern universities.

Early Work

New ideas, often but not always correct, proliferated in geology. The concepts of sedimentology however, advanced very little between 1915 and 1950.

During this time, many American and British sedimentary geologists devoted their time and effort to developing heavy-mineral and grain-size analysis techniques. While these techniques were then thought to constitute the mainstream of sedimentology (as can be seen, for instance, in the title of the *Journal of Sedimentary Petrology*, founded in 1931), their importance has since declined to a substudy of sedimentology. It was also during this time that the term *sedimentology* itself was coined and accepted, but only after being strongly opposed by some who disapproved of toying with the language (see Wadell 1931, 1933 for a defense of this new term).

The earliest study of the Hartford group from this era (Rice and Gregory 1906) echoed the arguments of the past and reiterated the estuarine concepts. These authors did, however, present one bit of reinterpretation that was to have a lasting effect even though it has recently been proved incorrect. After comparing plant fossils from the Virginia basins with those from "some European strata," Rice and Gregory (from Connecticut Wesleyan and Yale, respectively) concluded that the rift basins were filled with sediments of Keuper (Late Triassic) age.

This study followed the lead of studies by Newberry (1887, 1888) and Ward (1891) on the fossil fishes and paleobotany of the New York-Virginia basins—studies that had concluded that the beds were of latest Triassic ("Rhaetic") age. William C. Redfield's careful earlier work with the fish fossils, on the basis of which he assigned them to the Early Jurassic, was never even mentioned by Rice and Gregory. Their point of view in turn was seemingly corroborated by a 1911 study of the Connecticut fishes by C. R. Eastman of the University of Pittsburgh. Thus, by the turn of the century (and until recently), all the deposits in the rift basins of eastern North America were considered to be of Late Triassic age.

As noted by Sanders (1974, p. 7), this fit handily with the contemporaneous conception of local geologists that the "Palisades disturbance" definitively marked the end of the Triassic. This was the localized orogenic event that caused the postdepositional faulting in the rift basin (Rodgers 1967). Thus, the Triassic age designation of the rocks became widely accepted as much for its tidiness as for the scientific evidence supporting it.

Barrell: Models of Nonmarine Deposition

The driving force behind the reinterpretation of the Hartford group, until his untimely death in 1919 at the age of fifty, was Joseph Barrell. Barrell left a promising career in applied engineering geology to become professor of structural geology at Yale, where he was to build Yale's academic curriculum in structural geology.

Barrell had worked for the U.S. Geological Survey in Montana prior to going to Yale, and although his primary work there was with igneous rocks and processes, he had seen the western alluvial-fan and floodplain deposits firsthand. Even so, his general approach to geological problems, including those of the Hartford basin, was as a synthesizer of already-existing data; he did little field work himself and collected little new data.

Early Nonmarine Concepts

Barrell dealt with interpretive concepts of the Hartford group that had been primed for change by Davis. (When Lawson coined the term *fanglomerate* in 1913, he used the border-fault conglomerate of the eastern rift basins as one possible example.) The concepts were still relatively static, however, because most geologists still accepted the estuarine theory. In two series of papers, "Relative Geological Importance of Continental, Littoral, and Marine Sedimentations" (1906), and "Relations between Climate and Terrestrial Deposits" (1908), Barrell put nonmarine deposits into a much-needed perspective, emphasizing their previously neglected volumetric importance and diagnostic features. Published in a section called "Studies for Students" in the *Journal of Geology*, these papers undermined a century's worth of geological reasoning and interpretation in the Hartford basin.

The papers are long by today's standards: Barrell (1906) wrote forty-four pages, without figures, on the pros and cons of mud-crack interpretation, and concluded (overreacting, perhaps) that mud cracks were "one of the surest indicators" (p. 550) of a nonmarine depositional environment. This was entirely contrary to the tenets of the previous 100 years. Even Barrell, however, did not immediately put all of his ideas into practice. His 1910 paper with Loughlin on the lithology of Connecticut (a textbook for students more than an interpretive paper) took the noncommittal position that "opinions differ as to whether the valley was an estuary, a lake, or merely a broad plain receiving deposits of river waste" (Barrell and Loughlin 1910, p. 155).

Nonetheless, by 1915 Barrell had reinterpreted all of the basic evidence others had previously worked laboriously into variations of the estuary model. He eloquently demonstrated how previously known animal tracks, border conglomerates, desiccation cracks, fish fossils, and raindrop impressions fit much more neatly into a nonmarine environment of "migrating rivers or shifting lakes . . . [and] river flood plains of great breadth [that] were subjected to periodical drying . . . in mainly river deposits of an inland basin" (1915, p. 30).

Barrell reconstructed east–west cross sections of the Hartford basin (figure 3-1) using the concept of facies-equivalent lithologies that had been alluded to by Russell, Dana, and Emerson. Concurrently, Walther's 1894 ideas of

WEST

EAST

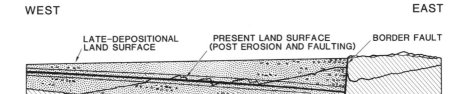

Figure 3-1. 1915: Barrell's new hypothesis of a syndepositional border fault and his concept of facies-equivalent lithologies. (After Barrell 1915, Figure 7, p. 29. Reproduced with the permission of the Connecticut State Geological and Natural History Survey)

facies were being promoted and advanced by A. W. Grabau in the United States, but—although these ideas had been used in Europe since the mid-1800s—they would not be widely accepted and used in England or America until the early to mid-1900s (Teichert 1958).

The sudden fitting together of all the pieces of the puzzle by means of the new depositional model must have been the source of almost as much relief to the abused geological logic (albeit on a more local scale) as was the later concept of plate tectonics. Like the early concepts of plate tectonics, however, the nonmarine model was only a framework upon which details had to be painstakingly added.

Very likely, part of what allowed Barrell the freedom to depart from estuarine ideas was the concept of Appalachia. This hypothetical continent, which supposedly was located offshore of the Atlantic coast and subsequently foundered, had been suggested to account for the voluminous sediments that obviously had been derived from an eastern source (but one that was no longer there) and shed westward to form the thick Paleozoic formations in the Appalachian region (Hall 1857). Its position would also have blocked access of any seaway to the Connecticut Valley, prohibiting the formation of an estuary. This concept is not altogether different in its paleogeographic implications from the presently accepted plate-tectonic continental reconstructions, although the mechanics of Appalachia's removal (through submergence) were purely hypothetical.

Origin of Redbeds

In addition to his studies of depositional environments, Barrell recognized that the red coloration of the rocks was principally a postdepositional, early diagenetic feature, caused by the "partial chemical decay" of the iron-bearing minerals. He compared the sediments (especially the unweathered labile feldspars) to those forming in semiarid parts of Spain and the southwestern United States. Finding the stage of diagenesis similar and compatible, he proposed that the syndepositional climate in Connecticut had been a semiarid one with hot summers and possibly cold winters. He envisioned deposition on a well-drained, oxidizing slope, rather than at the bottom of a low-lying basin.

Barrell based much of his climatic theory on two previously little-appreciated papers by Crosby (1885, 1891) that suggested that the postdepositional destruction of iron silicates released iron that was first oxidized and hydrated, and then—through the "spontaneous process" of dehydration—converted into hematite. The geochemical processes of this diagenetic sequence are now relatively well understood (Berner 1969), but in Barrell's time the description of the sequence was based entirely on empirical observation. Thus, according to Crosby, time was the only agent required to turn iron-bearing sediments red; additional heat was unnecessary, although Barrell suspected that the additional pressures produced by burial accelerated the process.

Russell had noted Crosby's 1885 paper but had dismissed it (prompting Crosby to write his second paper) because he felt that Crosby's initial reliance on solar heat for the dehydration process in the exposed soils was inadequate. Obligingly, Crosby downplayed the need for solar heat in his second paper, relying instead on a strictly spontaneous process and long periods of time.

Crosby's papers dealt with modern soils, however, and it is not clear whether his one reference to "the color of a deposit [being] . . . a function of its geologic age" (1891, p. 81) in fact referred to deposits of transported sediments or to the "sedentary" (weathered in place) saprolite soils he was primarily concerned with. Nonetheless, Barrell's mind was obviously primed to read the passage as being applicable to transported sedimentary deposits, and he proceeded to use Crosby's spontaneous process in his interpretations of the ancient sedimentary record. Unfortunately, Crosby's ideas and Barrell's use of them would go unaccepted by most geologists for half a century.

Synsedimentary Border Faulting, Geological Time

Although he failed to grasp the significance of the fine-grained organic-rich deposits, dismissing them as "swampy" deposits of minor importance, Barrell was the first to interpret correctly the tectonic and structural significance of the eastern border conglomerates. Using the evidence of the conglomerates, he inferred a syndepositional border fault, recognizing its influence on sedimentation (figure 3-1). As he wrote, "It is necessary to postulate a boundary consisting of a fault wall in order that renewed movements upon it may maintain such a long continued supply of coarse, yet local waste" (1915, p. 29). Previous ideas of the tectonics of the basin had been largely constrained by Dana's ideas of compressional tectonics and folding associated with a shrinking Earth; however, concepts of the extensional origin of rift valleys were still twenty-five years in the future.

Another subject Barrell addressed in depth was that of absolute geological time. In his 1915 article, he had reproduced Dana's speculative time scale that allotted only 9 million years (spanning the interval from 11 to 3 million years ago) to the entire Mesozoic era; 1915, however, was the same year that Arthur Holmes published his revolutionary paper on "Radioactivity and the Measurement of Geologic Time." Barrell was in print two years later with a long and insightful compendium that compared the estimates of geological time made by different quantitative methods. This study was instrumental in persuading the geological community to accept the radiometric absolute time scale (Hallam 1983).

Almost overnight, Barrell had overhauled the sedimentological interpretations of the Hartford group and laid the groundwork for detailed basin analysis and environmental studies. He had opened up the time scale, increasing the number of years allotted to the deposition of the basin fill by an entire order of magnitude. Had he lived, he might have continued to pioneer in such fields as sedimentology, although his later papers were tending toward more conceptual and broad-ranging topics such as the strength of the Earth's crust and isostasy. As it was, his death left a gap in the ongoing interpretative work in the Hartford group that the geological community was unprepared to fill. His immediate successors were uninspired in comparison to him.

Lull: Vertebrate Paleontology

Richard S. Lull, professor of vertebrate paleontology at Yale, reviewed the early and contemporaneous ideas on paleobotany, sedimentology, and vertebrate paleontology in the Hartford basin in 1915. Much of the review consisted of extensive passages quoted from earlier works, but he did point out the fact that paleontological evidence (mostly the trace-fossil vertebrate footprints) indicated a "profusion of life." This was a sticking point for many of the early models of the environment of deposition, both marine and nonmarine: marine models had to account for both an apparent absence of marine invertebrates and the presence of nonmarine vertebrates, whereas the analogue of a southwestern United States type of environment did not require an abundant paleofauna. Lull reemphasized that all of the paleontological evidence then available indicated freshwater to subaerial environments, providing strong support for Barrell's nonmarine interpretations.

An interesting speculation on Lull's part was that the first appearance of the footprints and vertebrate bones in the stratigraphic section of the Hartford basin may have marked the introduction of dinosaurs into the New World over a land bridge, and that this event may have marked the Triassic–Jurassic boundary. The source of this precontinental drift speculation may well have been a concept proposed by Charles Schuchert, Lull's contemporary at Yale. According to Hallam (1983), Schuchert postulated such land bridges as an alternative to Wegner's then controversial ideas on continental drift in order to explain faunal and floral similarities between widely separated continents. Given the lack of knowledge about the ocean floor at the time, long land bridges were not as outrageous an idea as they now seem. One had in fact been documented during the latest 1800s (see Hopkins 1967) across the shallow Bering Strait, in connection with the Pleistocene invasion of Asian fauna into the Americas.

Emerson: Holdover of the Estuarine Model

B. K. Emerson, who had written so graphically of tidal estuaries in Massachusetts in 1891 and 1898, was a somewhat flexible geologist, and when he published a U.S. Geological Survey Bulletin on the "Geology of Massachusetts and Rhode Island" in 1917, he tried to incorporate some of the new sedimentological concepts on nonmarine depositional environments. In fact, he elaborated on them, but not always successfully.

Emerson was aware of Barrell's alluvial-fan concepts, but he was led to different conclusions by his preconceptions as well as by several other observations. First, he had seen evidence for evaporites (figure 3-2) in the form of remnant gypsum and halite crystal molds, and for fresh water in the form of invertebrates. Both of these occur in the Chicopee Shale, a series of black lacustrine mudstones within the sandstone above the basalts. Second, he still believed that the basalt-flow time line could be extrapolated to prove that rocks exposed at the eastern and the western borders of the basin were deposited synchronously. Third, his correlation of conglomerate clasts with basement lithologies adjacent to the basin suggested significant north–south offsets of the two. Finally, he took the unweathered feldspar and rock fragments in the sediments to be evidence of arid, cold climates.

The final synthesis of Emerson's early estuarine tenets, his stratigraphic concepts, and this last bewildering set of observations, was a Rube Goldberg

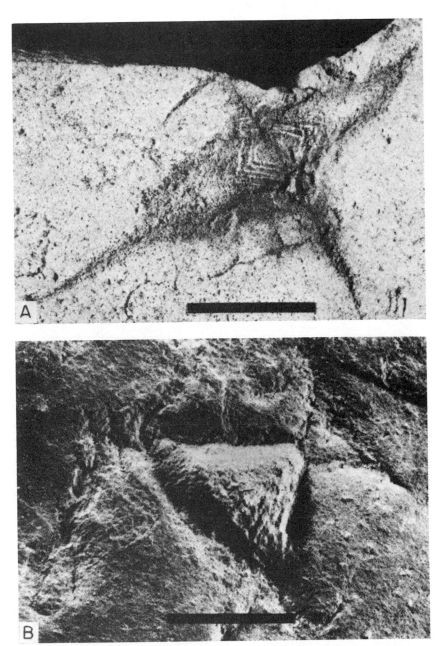

Figure 3.2. Pseudomorphs of halite crystals in the lacustrine "Chicopee Shale" facies of the Portland Arkose. Lake-bottom waters were highly saline, and possibly anoxic as well. Bars are 1 centimeter long. (Reprinted from Parnell 1983, Figure 3, with the permission of *Sedimentology*)

machine of sorts, in which glacial, lacustrine, flood, and tidal events were mixed to produce the final sedimentary deposits. As set out by Emerson (1917, p. 132) the first stage of deposition was marked by floods and avalanches that produced the border conglomerates in an arid subglacial environment. Subsequently, the climate became more humid, a lake developed, and the Sugarloaf Arkose (mapped as a synchronous deposit on both the east and the west sides of the basin) was deposited along the "shore," while the Longmeadow Sandstone was deposited "offshore" and the fine-grained Chicopee Shale was deposited in the central, deep-water positions as the waters "rose" to their "maximum width." With time, the lake was made shallower by the extrusion of basalts "oozing along the bottom of the bay" (p. 92), and ensuing desiccation of the lake produced evaporites. Finally, a large mud flat was produced by renewed flooding. Somewhere within the time represented by this model, "strong northward [tidal] currents" offset conglomerate clasts from their immediate source rock areas along the edge of the basin.

The model was not terribly elegant, but at least it represented an effort to fit all the existing field data into a logical depositional environment and sequence of events. Its principal failure was in not recognizing the eastern Sugarloaf Arkose as a distinct and chronologically different deposit from similarly named deposits exposed on the western edge of the basin. The subdivision of these two areas—and the extension of the stratigraphic ideas deduced from the well-defined formations in Connecticut—would not occur until the mapping of Willard in the early 1950s. Willard would also rename and standardize the formations, eliminating the confusion created in the Deerfield basin where Emerson labeled a mapped unit as Longmeadow Sandstone in 1898, but as Chicopee Shale in 1917.

Longwell, Russell, and Foye: Sedimentology of Fan Deposits and Proof of the Eastern Border Fault

Geologists generally accepted Barrell's models of depositional environment (aside from climate) throughout the remainder of the early 1900s. H. G. Reading (1978) suggests that the development of sedimentological ideas essentially stopped during these years; and except for investigations into the origin of the red color of the sediments, this is true of work in the Hartford basin.

In other fields, however (principally that of structural geology), progress continued in the Hartford basin. Many of the discoveries here had significant implications for the sedimentological interpretations, and the sedimentology in turn was used to support the structural concepts.

Basin Structure

Three papers published in 1922 dealt with the eastern border fault and the structural history of the basin. The two main models discussed were Davis's unfaulted, downwarped basin with postdepositional faulting, and Barrell's synsedimentarily faulted half-graben configuration. All three papers concluded that Barrell was correct. W. G. Foye (1922) from Connecticut Wesleyan listed the general evidence, summarized as follows:

1. Streams in a downwarped basin would have been incompetent to transport the large clasts found in the basin fill.
2. Sediment grain size was different on either side of the basin; therefore, the two basin-margin structures must have been different. (Nowhere was the obvious westerly predominance of crossbedding noted.)
3. The thinner strata in the Pomperaug outlier suggested a westward-thinning wedge of sediments, and the overall decrease of structural dip from west to east noted by Davis could be interpreted as synsedimentary faulting and subsidence preferentially on the east side of the basin.
4. Clast sources for the conglomerates are often immediately east of the border fault—areas that should have been buried during the Triassic according to Davis's downwarp model.
5. The very coarse eastern border fanglomerates were indicative of border faults.
6. There was evidence of synsedimentary volcanism along the inferred faults.

C. R. Longwell (of Yale) and W. L. Russell also published papers in 1922, adding specific data to Foye's observations. Russell noted that basalt cobbles were present in the conglomerates above and between the flows, indicating synsedimentary faulting. More importantly, he observed that a thick quartz vein paralleled and lined the eastern border fault in several areas, and that quartz clasts attributable to this vein were present in the conglomerates. From this, he inferred that the fault had been present prior to the Triassic and that it had also been active during the filling of the basin.

Alluvial-fan Deposits and Facies Concepts

Both Russell and Longwell noted that the basalt flows often wedged out against the conglomerates, evidence supporting the concept of alluvial fans with inclined surfaces along the eastern border fault. Russell even speculated that the slope of the fan might have been several hundred feet per mile, but Longwell was the first to look at the sedimentary structures of the conglomerates. Longwell, who had worked in the Basin and Range province in Nevada, presented four pieces of sedimentological evidence for alluvial-fan deposits, summarized as follows:

1. The conglomerates are lenticular, thinning laterally and grading into finer lithologies.
2. The internal bedding is "rude" and lacks sorting.
3. The clasts are angular.
4. There is a point source for many of the clasts in the adjacent basement rocks.

In 1928, Longwell was called upon to write a rebuttal to J. K. Roberts, who had not yet grasped the concepts of facies and of Walther's law. Roberts (1928), still in the strong grip of layer-cake geology, had interpreted the Virginia rift basins as having been concentrically filled by successively finer-grained layers of sediments and had extrapolated that antiquated model to the Hartford group (figure 3-3).

Longwell later wrote a generalized guide (1933) in which he inferred environments of floodplains (on which mud cracks, tracks, and raindrop impressions were preserved) and temporally equivalent alluvial fans, swamps, and shallow lakes. According to Longwell, the black mudstone beds were deposited in the lakes during persistent "exceptionally moist conditions," but he presented no new sedimentological evidence or arguments to support his interpretations.

Longwell's final paper on the subject (1937) used the Hartford group as an example of the concept of sedimentation controlled by tectonics, a popular subject of that era. For Longwell, the conglomerates constituted prima facie evidence for local, large-scale faulting. He also wrote (p. 437) of "mud flow" deposits, whose characteristics were clasts with a "pell mell arrangement in a fine-grained matrix."

Thus, Longwell had begun to codify the sedimentary characteristics of the Hartford group's alluvial-fan deposits, even though they were not his

WEST   EAST

Figure 3-3. 1928: "Layer-cake" interpretation of the lithofacies of the Hartford group, a holdover from the era when many American geologists discounted the concepts of facies. (After Roberts 1928, Figure 15, p. 169. Reproduced with the permission of the Virginia Division of Mineral Resources)

primary concern. His lead was followed by Reynolds and Leavitt (1927) in a short paper that described the coarse, thick, angular portions of the Mount Toby conglomerate in Massachusetts as a scree or talus against a high-relief fault.

Foye and Pillow Basalts

Emerson (1898*b*) had described "spherical blocks" in basalts in Massachusetts that suggested to him a partial remelting of lava, but he had no concept of their true origin. Pillow basalts had only just been recognized in Hawaii and correctly interpreted as a subaqueous phenomenon at around the turn of the century (Lewis 1914). Controversy in applying a subaqueous conclusion to these structures in the ancient record, however, remained—especially as to what were now considered nonmarine deposits. In 1924, Foye described the first pillow basalts noted as such in the Hartford basin (figure 3-4), but under the influence of the prevailing subaerial concept of the environment and in view of the seeming absence of underlying subaqueous sediments, he interpreted them as a postdepositional feature.

Figure 3-4. Pillows in the lower part of the Talcott Basalt. Such structures had been noted earlier but were not interpreted until 1924. Pillows are the chilled skin and resulting internal structure of lobes of lava that advanced into bodies of water. (Photo by the author)

Hubert et al. (1978) have since suggested that a lake may have been present at the time of extrusion of the basalt, but that it was so new that it lacked a sedimentary record. Thin, recently deposited lacustrine sediments might also have been destroyed by being incorporated into the volcanic/sedimentary agglomerate that commonly underlies the pillowed zone.

Bain and Wheeler: More Observations on Basin Structure, with Asides on Sedimentology

The structure of the Hartford basin continued to be a topic of debate in the early half of this century, as varied and often conflicting evidence was periodically turned up by different workers in different parts of the basin. Nevertheless, some observations were made on the sedimentary aspects of the local rocks as well. Bain (1932) described lobes or tongues of conglomerate of the Mount Toby Formation that had prograded out onto lacustrine shales. Basal fluting and the synsedimentary distortion of the mudstones below and around the large, unsorted, and angular conglomerates suggested to Bain that these might have been "mud flows or landslides" that had been introduced directly into the lacustrine environment.

Bain also disagreed with the talus or scree interpretation of Reynolds and Leavitt (1927), noting oriented clasts, slight water-transport rounding, and intercalated shale lenses within these Mount Toby conglomerates. He inferred from these data that the deposits were "wash accumulations" rather than rockfall screes.

With regard to the lacustrine shales, Bain (1932, p. 68) cited another geologist as being of the opinion that the associated crystal casts that Emerson had taken to be halite bore "greater resemblance to frost crystals around desert playas." (Krynine in 1950 *re*-reinterpreted these as gypsum crystal casts.) Bain proceeded to infer a cold, arid climate, complete with glacial highlands. He would later (1957) refer to these deposits as a "bolson mud flat facies" with evaporites and flash floods, while attributing the black mudstones to local small ponds; but he offerred no new evidence for his environmental interpretations.

Another geologist, Wheeler (1937), put together a well-documented suite of evidence that showed the variable nature (alternately faulted or onlapped) of the western margin of the basin. He then extended his lines of reasoning onto tenuous ground and revived Russell's broad-terrane hypothesis. Much of his reasoning was nebulous, and some of it was simply wrong. He suggested that, since the narrow estuarine model was no longer in vogue and since there was no continuous western border fault, the original western extent of the deposits could have been essentially unlimited. He also dismissed the wedging-out relationships shown by the thinner strata of the Pomperaug outlier on the grounds that correlations were inexact in these unfossiliferous deposits—ignoring the correlatable basalts.

Redbed Controversy

Although most studies were focused on the structural aspects of the Hartford basin during this time, the subjects of sedimentology and environments of deposition were still being considered. Aspects of geochemistry and diagenesis however—specifically, with regard to the origin of redbeds—had not progressed significantly beyond Russell's ideas of the mid- to late 1800s, despite Barrell's cogent arguments in the first decades of the 1900s.

Goaded by Barrell's concept of an arid climate, Dorsey (1926) and Raymond (1927*a*) wrote papers defending the idea of redbeds as indicators of nonmarine, warm, humid environments. As Russell had, both Dorsey and Raymond confused the development of red soils in source areas with the deposition of red sediments in basins. Both authors acknowledged Barrell's contention that the dehydration of iron hydroxides and the dissolution of iron silicate minerals required significant amounts of time, but curiously, they allowed for this lengthy process only in potential source areas. Thick, red, lateritic soils were envisioned as accumulating over long periods of time in warm, humid, tropical climates, and then as being rapidly dumped into and buried in nearby basins. This was exactly Raymond's interpretation of the red sediments of the Hartford group (1927*a*, p. 247). Rapid burial was envisioned as a necessity for the retention of the red color. On the other hand, long-term processes of iron dehydration were denied in areas of sediment accumulation. The argument was that modern deserts were never red because no known "decomposition" (diagenesis) occurred there, and therefore that they could not have been the environments of deposition for ancient redbeds as Barrell had suggested.

Glock (1927) published a rebuttal to Raymond and Dorsey in defense of the arid redbed concepts, but again knowledge of the geochemistry of iron oxides was insufficient to uphold an effective argument. Glock's rebuttal centered on the great volumes of sediment found in the basins of the southwestern United States—sediment that had obviously been produced under conditions of very little rainfall. Therefore, he suggested, the large volumes of sediment in the rift basins could not by themselves be used to prove the previous existence of an environment with abundant water supplied by a humid climate.

These studies were part of a continuum of papers written about redbeds. Few of the studies directly concerned the strata of the east coast rift-basins, but often enough these rocks and their interpretations were cited as evidence for one theory or another. Despite all of these studies, a large mental step was required to go from thinking in terms of primary red soils, with their obvious modern analogues, to thinking in terms of the secondary production of red colors in postdepositional diagenetic environments, where the mechanisms were not readily apparent.

For the time being, the redbed problem remained unresolved because of insufficient data. The warm-humid concept could produce voluminous red sediment, but it suffered from the lack of a modern environment in which such sediments were known to be accumulating. The opposing arid-climate camp had what was felt to be the proper modern environment, deposits of which were similar in most respects to those of the ancient—except that none of the accumulating sediments were known to be red.

Krynine: Petrography, Climate, and Redbeds

Sorby's ideas and methods for petrographic examination of rocks were developed in the late 1800s but were only beginning to receive widespread acceptance and use in the 1920s and 1930s. The prime advocate and user of petrographic techniques in the United States was Paul D. Krynine, a colorful figure—both aggressive and self-assured—with a Ph.D. from Yale and a professorship at Pennsylvania State University.

Although Krynine's paper on the "Petrology, Stratigraphy, and Origin of the Triassic Sedimentary Rocks of Connecticut" was essentially completed by 1936 as his Ph.D. dissertation, its publication was delayed until 1950 by World War II. In the paper, Krynine included exhaustive petrographic descriptions and characterizations of the sediments. His significant and lasting contributions were in other aspects of the geology, however; much of the petrography added little to the existing understanding of the origin or conditions of deposition of the rocks of the Hartford group. Pettijohn (1984, p. 240) writes that this was an era when sedimentologists, equipped with elaborate laboratory analysis of textural parameters of clastic sediments, were under "the illusion that to quantify the subject was to make it a science"; but the numbers were in fact rarely related to interpretations or predictions.

Although Krynine's study was exhaustive, the aspects of geology it did not address are revealing. The study did not use even the modest level of knowledge possessed in the 1930s about primary sedimentary structures, and determinations of paleoflow through the measurement of crossbeds would not come into general use until almost a decade after Krynine published. Nevertheless, Krynine did make several significant contributions to the understanding of the Hartford group.

Stratigraphy

One of the most useful things Krynine did was to begin giving formal stratigraphic names to the formations in Connecticut (table 3-1). Most of the formations were still being referred to by the unwieldy, geographically descriptive terms of Percival's 1842 work. Krynine's study included a detailed geology map of the Triassic–Jurassic of central and southern Connecticut (figure 3-5). The mapped categories, however, were a curious blend of stratigraphy, based on petrographic mineral assemblages, and lithofacies. Thus, the newly named New Haven Arkose in southern Connecticut was divided into upper and lower stratigraphic divisions (based on different trace-mineral content) and a basal arkose, whereas the same formation in central Connecticut was divided into a "redstone" facies and an "interfingering zone of redstone and arkose"—a purely lithologic division. These subdivisions have not been used by later geologists.

The Portland Arkose was mapped more consistently as those rocks above the last basalt, the only variation being the basin-margin fanglomerate lithofacies. Four small areas of distinctively different lithology were also mapped: "Dark Shale" and "Limestone." The significance and lateral extent of these rocks was not recognized by Krynine, and little was made of them for several decades.

In another area, Krynine did a calculation on the syndepositional extent of the basin to the west, geometrically extrapolating from the thick, main-basin stratigraphy, through the thinner units of the Pomperaug outlier, to a proposed feather edge of the basin about 3½ kilometers west of the outlier

Table 3-1. **Correlation of the different stratigraphic units used by different authors in the Hartford basin.**

AUTHOR	Percival 1842	Davis 1898	Krynine 1950	Lehman 1959
	EASTERN SANDSTONE	UPPER SANDSTONE	PORTLAND ARKOSE	PORTLAND ARKOSE
	POSTERIOR TRAP	POSTERIOR TRAP	UPPER FLOW	HAMPDEN BASALT
FORMATION	MIDDLE SHALE AND TRAP	POSTERIOR SHALE / MAIN TRAP / ANTERIOR SHALE	MERIDEN FM AND MIDDLE FLOW	EAST BERLIN FM / HOLYOKE BASALT / SHUTTLE MEADOW FM
	ANTERIOR TRAP	ANTERIOR TRAP	LOWER FLOW	TALCOTT BASALT
	WESTERN SANDSTONE	LOWER SANDSTONE	NEW HAVEN ARKOSE	NEW HAVEN ARKOSE

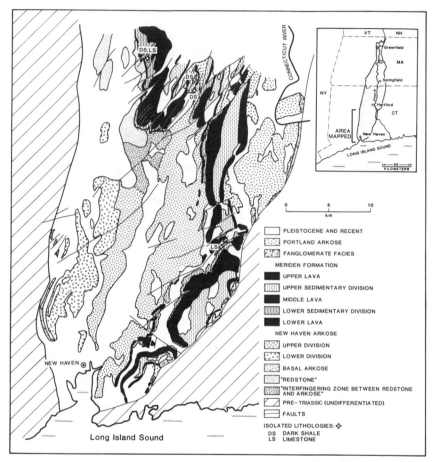

Figure 3-5. 1950: Krynine's geological map of the southern end of the Hartford basin. Krynine used a somewhat confusing scheme of mapping that mixed lithologic and statigraphic classifications. He also did not extrapolate lithologies under major areas of cover. (Modified by the deletion of intrusive basalts; after Krynine 1950. Reproduced with the permission of the Connecticut State Geological and Natural History Survey)

(figure 3-6). Such an approach had also been undertaken by Foye (1922), but it lacked the mineralogical control on stratigraphic correlation.

Uses of Petrography

Despite the tedious descriptions of grain-size distributions and fabrics, Krynine's petrographic studies did add to the knowledge of the strata in several areas. The mineralogy of the rocks allowed him to conclude that

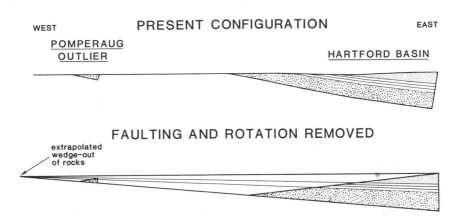

Figure 3-6. Krynine argued that the westward thinning of formations disproved the broad-terrane hypothesis, and that the strata of the Hartford basin had originally wedged out just west of the Pomperaug outlier. (After Krynine 1950, Figure 33, p. 122. Reproduced with the permission of the Connecticut State Geological and Natural History Survey)

there was no significant contribution of sediments from western sources— that, in fact, even the sediments of the Pomperaug outlier were probably derived from sources east of the main basin's eastern border fault. In certain cases, Krynine was able to decipher the sequence of erosion in the source area by demonstrating an inverted metamorphic-to-granitic compositional sequence within the sediments, reflecting the unroofing sequence of source rocks. Such vertical distributions allowed him to subdivide some of the formations, but to date, these divisions have had little application to sedimentological problems.

Krynine also used his suite of over twenty trace or heavy minerals for purposes of east–west stratigraphic correlation, documenting that the bulk of the sediments in the Pomperaug outlier correlated with the New Haven Arkose. Another important use of the heavy-mineral study was in document-ing primarily east-to-west paleodrainage. Mineral assemblages from a known source terrane can be traced westward out into the sediments of the basin, but adjacent assemblages do not seem to be mixed until as far west as the Pomperaug outlier, suggesting that little or no north–south longitudinal drainage occurred within the preserved part of the basin (figure 3-7).

Redbeds and Climate

The rare interpretations of depositional environment that Krynine made were closely tied up with his theories on climate and the origin of the redbeds. Krynine had spent time previously in southern Mexico working for an oil company, where he observed what he thought were deposits containing

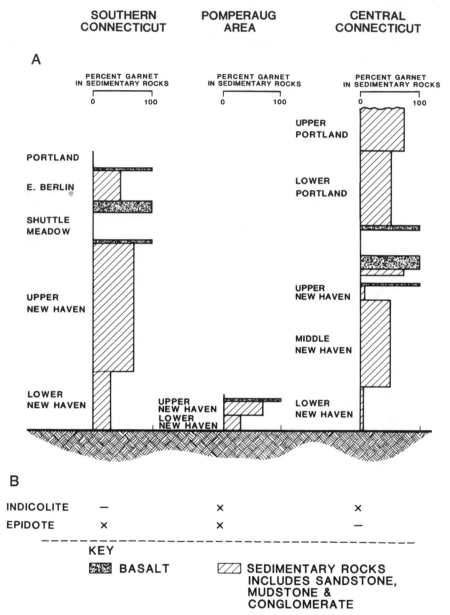

Figure 3-7. Krynine used heavy and trace mineral assemblages (A) to subdivide and correlate formations, and (B) to infer paleogeography: the lack of north–south overlap of trace mineral suites from the eastern highlands provinces indicated that paleodrainage was primarily east–west. Overlap did not occur until as far west as the Pomperaug outlier. (After Krynine 1950, Figure 6, p. 36. Reproduced with the permission of the Connecticut State Geological and Natural History Survey)

both fresh feldspars and red sediments—deposits that were forming under conditions of high relief and heavy precipitation in a humid, tropical setting. Krynine published this work (1935), but his "modern" red deposits are actually late Tertiary strata that he had incorrectly ascribed to (or remembered as) a modern floodplain environment (Hubert 1986). The modern deposits in this area are in fact yellow-brown, not red.

This inaccurate interpretation gave Krynine a supposedly modern environment in which red sediments were accumulating to use as an analogue to the ancient conditions. Like others before and since, he subsequently interpreted much of what he saw in the ancient in terms of what he was most familiar with in the "modern."

The concept of the formation of red sediments primarily in a humid sediment source area was therefore a dominant factor in Krynine's thinking. He stated categorically (1950, p. 149) that "If the present is the key to the past, one thing appears to be certain: that is that widespread primary red color in sediments is prima facie evidence for a warm, humid (preferably tropical or subtropical) climate in the source region from which these sediments came." Primary red sediments were envisioned as having been deposited rapidly on a Barrellian well-drained slope in "nonreducing" conditions, which preserved the primary red color.

His supporting arguments for this statement differed little from Russell's, Raymond's, and Dorsey's of years before, but Krynine did offer one ambiguous observation as proof of the primary red color of the sediments. He noted an arkose below a basalt flow that had been "bleached by the thermal action of the lava, thus conclusively proving that the red color of the Triassic beds is of primary, prediagenetic origin" (1950, p. 65). This statement neglects the possibility that such contact metamorphism might alter the local chemistry and permeability, affecting any subsequent diagenetic processes.

Again, no red soils were known to be forming in modern deserts, and a lateritic source area required a warm, humid climate; thus, a warm, humid paleoclimate was postulated. Such a climate was apparently supported by the widespread lacustrine deposits of organic mudstones in the middle of the formation, which indicated to Krynine either structural damming of drainage or rates of high precipitation and warm temperatures. This, plus the necessity of providing unweathered feldspars from an area of high-intensity weathering, led to his compromise model: a highland mantled with thick red lateritic soils was incised by streams that eroded into fresh bedrock— a concept mentioned briefly by Raymond (1927b)—and the resulting sediments from soils and bedrock were mixed and deposited rapidly in the adjacent basin.

With this preliminary model in mind, Krynine next had to account for the sedimentological evidences of aridity—primarily mud cracks and salt casts.

He argued long and hard to demonstrate examples of the formation of such phenomena in seasonally humid climates, before eventually proposing a "savanna" type of climatic regime. Thus, his final climatic model was warm, tropical, and humid ("above 50 inches" of precipitation) with a marked dry season (up to three months long). This model was accepted for years, until definitive work on the formation of redbeds was undertaken by Walker and Van Houten in the 1960s and early 1970s.

Krynine's Legacy

Although Krynine suggested in his introductory material that the study of facies relationships between the different lithologies would be rewarding, it received little attention in the report. This, together with his nebulous conceptions of the details of depositional environments, was overshadowed by the mass of petrographic detail presented. Although petrography proved to be of value in deciphering the stratigraphy and structural history of the basin, it was usually of little help in understanding the sedimentology of the rocks. Nonetheless, Krynine's environmental interpretations stuck, if only because of the sheer volume of data he presented, and the next attempts at sedimentological interpretation were not made until the late 1960s.

Krynine made another use of the petrography of the sediments in the Hartford basin. In 1942, he published a paper that suggested that the mineralogical composition of sandstone in a basin was a function of whatever stage of the "diastrophic" or geosynclinal cycle was concurrent with deposition of the sandstone. This study used formations from the Appalachian area as examples of preorogenic quiescent stages (orthoquartzites) and of the subsequent active geosynclinal stages (graywackes and second-cycle orthoquartzites), while the Hartford group provided the basis for continuing the concept: arkosic compositions were proposed to be typical of orogenic and postorogenic phases.

The idea that petrographic composition could be directly related to tectonics was popular for a few decades in the middle of this century, corresponding to the heyday of both geosynclinal theory and petrology. On the basis of these two concepts, it was often stated that the essential points of a basin's history could be deduced from a single thin section. This maxim began to lose adherence when, among other things, Klein (1962) pointed out that the Triassic–Jurassic sediments of Nova Scotia and New Brunswick, deposited during the same tectonic stage as those in Connecticut, did not fit the compositional (arkosic) pattern. Klein suggested that the source area, the provenance of a sandstone, provided the primary control on its composition—an intuitive and simple idea that had been overshadowed by theory run wild.

4 / 1950 to the Present

As chronicled by Hsü (1973) and McBride (1973), detailed knowledge and more sophisticated concepts of sedimentology became prevalent only after World War II. This followed from detailed studies of modern depositional environments, of sedimentary processes, and of environment-specific interpretations of primary sedimentary structures. Such studies were spurred in large part by the petroleum industry's search for oil.

The expansion of the science of sedimentology meant that, henceforth, most geologists would find more than enough to study within a single formation of the Hartford group, rather than attempting the previous approach of addressing the whole basin-fill sequence. As interpretations of ancient depositional environments were made to confront and use the abundant evidence of sedimentary structures, better-reasoned and more fully documented geological conclusions began to appear in print.

Sedimentary petrography was no longer the main thrust of sedimentology by 1960. The concepts and techniques of basin analysis (as introduced by Potter and Pettijohn in 1963) had taken away much of its prominence in geological methodology. Potter and Pettijohn showed how sedimentary structures can be useful in reconstructing paleoflow regimes and paleogeography, and how they need not just be noted descriptively.

The concepts of structure also underwent a revolution with the rediscovery and documentation of plate tectonics in the late 1960s. Rift basins with the form of grabens and half-grabens were now seen as having resulted from extensional stresses associated with the rifting of continents, instead of being force-fit into compressional regimes related to the Appalachian mountain system. The mobility of continents did away with the contrived concept of Appalachia—allowable in previous years when geologists had total ignorance of the character of the sea floor—and replaced it with Africa, in a more geologically coherent scheme. Finally, detailed geological quadrangle mapping was encouraged at this time (Bell 1985), and specific formation and facies distributions began to be mapped.

Writing about the history of geology becomes increasingly difficult as the present day is approached. Many geologists now studying these deposits may not be happy to have their ideas examined historically, especially since there is as yet not much of a historical framework for them. Moreover, many ideas are still in a state of flux, and it is not possible to tell which ones will hold up to the tests of time and accumulating data. Nonetheless, this book would not be complete without coverage of the present stage of geological thought, and this chapter is therefore offered as an attempt to explain the concepts and interpretations that are currently being applied to the basin.

Mappers and Stratigraphers

The first order of business for this half of the century was the partial reconciliation of the stratigraphy and its nomenclature between Massachusetts and Connecticut. This was started by Willard (1951, 1952), who corrected the inconsistencies in Emerson's 1898 and 1917 maps by restricting the term *Sugarloaf* to the sandstones below the Deerfield lava in the Deerfield basin and by renaming the sandy sediments above the lava as the Turners Falls Sandstone (figure 4-1).

The terms *Longmeadow Sandstone* and *Chicopee Shale* in the northern Hartford basin were eventually lost, as Willard's stratigraphy was extended south and Krynine's stratigraphy was extended north. Colton and Hartshorn (1966) even mapped a quadrangle that straddled the Connecticut–Massachusetts border, using the nomenclature from Connecticut.

The Chicopee Shale is lithologically and environmentally similar, but not stratigraphically equivalent, to the organic, fine-grained lacustrine deposits between the lava flows in Connecticut. Its loss as a formation obscures some of the facies relationships, but the revision of the Sugarloaf was needed. Previously, Sugarloaf Arkose had occurred at two separate stratigraphic levels, separated by two sedimentary formations and a basalt flow.

Figure 4-2 shows the rocks as they were correlated by Reeside et al. (1957). The recent geological map of Massachusetts (Zen 1983; Robinson and Luttrell 1985) retains a few of the old lithologic distinctions within the mapped time-rock stratigraphic units (figure 4-3).

As reported in Reeside et al. (1957) and in Rodgers, Gates, and Rosenfeld (1959), ongoing work through the 1940s and 1950s on dinosaur, fish, and plant fossils continued to support a Late Triassic age for these deposits. Reeside et al. (1957) did put the uppermost sediments(the Portland Arkose

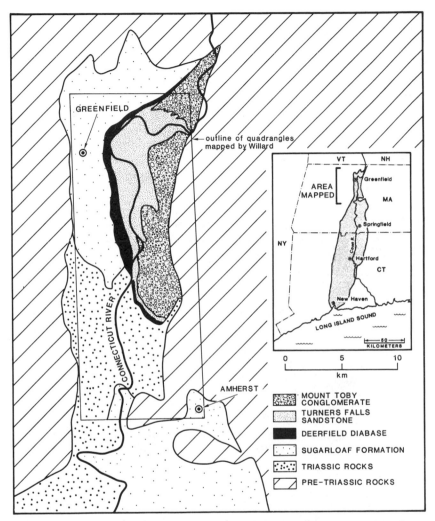

Figure 4-1. Willard's 1951/1952 restructuring of the geological map and stratigraphic nomenclature of the Deerfield basin. Willard also distinguished and mapped three lithologic subdivisions of the Mount Toby Formation that are not portrayed here for lack of space, and he mapped the strata as Triassic in age. (After Willard 1951; and Willard 1952. Reproduced with the permission of the U.S. Geological Survey)

and its equivalents) into the Rhaetian, a stage of somewhat ambiguous affinities usually thought of as the last Triassic stage.

The final paper of this decade of the 1950s, other than a series of pertinent quadrangle maps, concerned the discovery of another in-faulted outlier of arkosic sediments west of the Hartford basin (Platt 1957). Exceptionally small, it was estimated to have an areal extent of only 0.8 square kilometers, but it was nonetheless grandiosely named the Cherry Valley basin (see figure I-3). It was revealed only after some of its post-Triassic cover had been washed away by disastrous flooding in 1955. Both the sediments and the structure of this deposit mimic those of the Hartford basin and the Pomperaug outlier.

By the end of this time, Krynine's Meriden Formation had been given group status, and the sedimentary units within it subdivided by Lehman

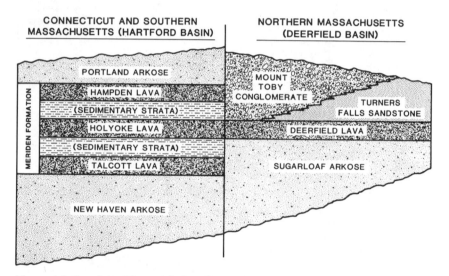

Figure 4-2. Stratigraphic correlations between the Hartford and Deerfield basins, as presented by Reeside et al. 1957.

Figure 4-3. 1983, 1985 *(facing page)*: Current geological maps of the Hartford and Deerfield basins. This adaptation shows most of the faults; for simplicity, all igneous rocks are depicted with the same symbol, and none of the intrusives are portrayed. Note that the map units of Massachusetts still retain some of their lithologic distinctions; thus, some discrepancy remains across the state line with Connecticut. Emerson's "Chicopee Shale" no longer has a mapped expression. Only the lowest rock unit below the basalt flows is unbroken between the Deerfield and Hartford basins. This unit is called the New Haven Arkose and the Sugarloaf Arkose in the Hartford and Deerfield basins, respectively, but no dividing line is mapped between them. Note also that most of the faults in the basin are still only well defined where they cross the erosion-resistant basalts. (After Rodgers 1985; and Zen 1983. Adapted and reproduced with the permission of the U.S. Geological Survey, the Connecticut Geological and Natural History Survey, and the Massachusetts Dept. of Public Works)

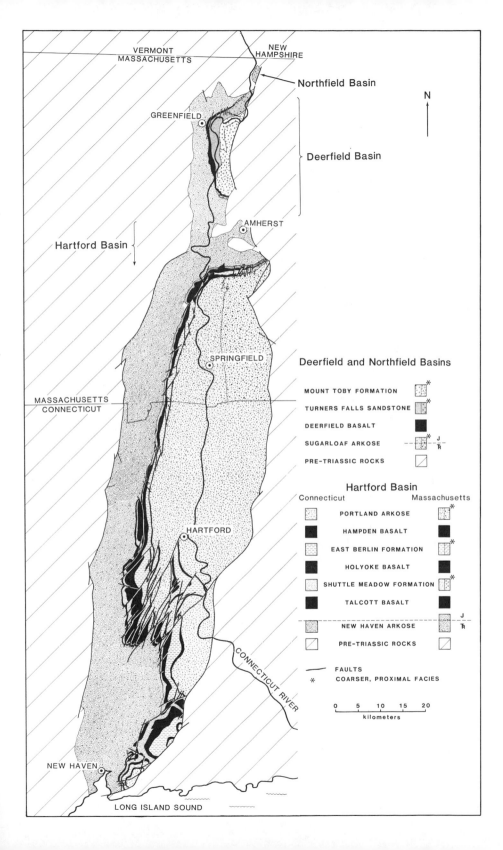

VERMONT / MASSACHUSETTS

NEW HAMPSHIRE

Northfield Basin

N

GREENFIELD

Deerfield Basin

AMHERST

Hartford Basin

SPRINGFIELD

MASSACHUSETTS
CONNECTICUT

HARTFORD

CONNECTICUT RIVER

NEW HAVEN

LONG ISLAND SOUND

Deerfield and Northfield Basins

MOUNT TOBY FORMATION *

TURNERS FALLS SANDSTONE *

DEERFIELD BASALT

SUGARLOAF ARKOSE * $\frac{J}{\text{Tr}}$

PRE-TRIASSIC ROCKS

Hartford Basin

Connecticut   Massachusetts

PORTLAND ARKOSE *

HAMPDEN BASALT

EAST BERLIN FORMATION *

HOLYOKE BASALT

SHUTTLE MEADOW FORMATION *

TALCOTT BASALT

$\frac{J}{\text{Tr}}$

NEW HAVEN ARKOSE

PRE-TRIASSIC ROCKS

FAULTS

* COARSER, PROXIMAL FACIES

0   5   10   15   20
kilometers

(1959) into the Shuttle Meadow Formation below and the East Berlin Formation above the central Holyoke Basalt. This produced the stratigraphic nomenclature currently in use in Connecticut.

Miscellaneous Studies of the 1960s: Structure and Basalt

Structure

J. E. Sanders, professor of geology at Yale, wrote a number of papers on the Hartford basin between 1960 and 1974. Although his later papers used detailed sedimentological concepts, his early papers (1960, 1963) dealt entirely with the structure of the basin. He and his co-workers discovered a previously unmapped, postdepositional fault that accounted for some of the ambiguous folds in the southeastern part of the basin, and his 1963 paper reconstructed the syndepositional and early postdepositional structural history of the basin.

In this 1963 paper, he also put forward the first ideas that the elongate basins of eastern North America may in fact have been of extensional, rifted origin. In a guide book he edited (1957), Bain had previously suggested that the East African rifts were analogous to the Hartford basin, but he was under the impression that there was evidence for wrench faulting and compressional thrusting in both areas, and that this explained the origin of these basins. Sanders also referred to the East African rift valleys as analogies, but in his other brief reference to present-day rifts in the paper he was hampered by the as-yet incomplete knowledge of the ocean floor and oceanic crust. He speculated that there may have been a "circumglobal" Triassic fracture zone— which was not far from the mark, in light of the presently known, extensive Atlantic-margin rift-basin system—but the idea was incorrect in that he cited Heezen's (1960) work on the oceanic Mid-Atlantic Ridge and rift system as the analogue.

Sanders (1960, 1963) also resurrected Russell's broad-terrane hypothesis on the basis of structural data. He saw the structural symmetry in both the basin-fill sediments and the regional basement rocks between the Connecticut and New Jersey areas, as well as their apparently similar stratigraphic history, and concluded that the broad-terrane concept was viable. Few have followed his lead to date.

Basalt

Several studies on the character and origin of the basalt flows were published in the 1960s. Chapman (1965) did detailed petrography and outcrop studies of the Hampden Basalt, concluding that it originated from fissures on the western edge of the basin and flowed northeast. This paleogeographically significant observation has not yet been used by others. Chapman recognized three extensive flows within the Hampden Basalt (locally up to eight less-extensive flows) that were amalgamated to form the unit.

One of Chapman's outcrops was the same extensive quarry face that Silliman had examined. The quarry that was "nearly three miles S.S.W. from the city of Hartford" (Silliman 1830, p. 122) in 1830, is now inactive and has become Rock Ridge Park on the western edge of the Trinity College campus, well within the Hartford urban area. Foose, Rytuba, and Sheridan (1968) also recognized northeasterly flow in tongues of basalt in central Massachusetts. They were able to locate the vent and to document the extent of five small lava flows originating from it.

Papers by de Boer (1967, 1968a, 1968b) significantly advanced understanding of the regional tectonic and stress regimes but were notable for having been published just prior to the general acceptance of plate tectonics. Several of the inferences de Boer made from his data stood at the brink of being plate-tectonic schemes, but they were too early to benefit from the implications of plate-tectonic theory. From his data, de Boer suggested that there had been "a progressive northeastward expansion of the broad geanticlinal arching of the Appalachians in the early Mesozoic time" (1968b, p. 609), and that volcanic activity had concurrently shifted in that direction. His paleomagnetic data also puzzled him, in that it suggested an accelerated and relatively rapid shift in pole position during the Late Triassic–Early Jurassic. The extensional stresses that caused the rift basin to form, as well as the broad arching and the shift in volcanic activity, would soon be integrated into a scheme of continental breakup and drift, but at the time of de Boer's papers they could only be reported as apparently unrelated facts.

The mixed-age interpretations of 1967–68 offer an interesting example of logical reasoning that led to possibly inaccurate conclusions because of the underlying assumptions. The paleomagnetic data collected from the dikes by de Boer (1967) showed an apparent pole position that was not coincident with the known Late Triassic pole positions, suggesting that the basaltic dikes of the Appalachian region were of Jurassic age. This was corroborated by the few radiometric age dates then available, and it is in fact the currently accepted age designation for the dikes. The rift-basin strata were still thought to be of Late Triassic age in 1967, however, and therefore the intrusive igneous rocks were assigned to a postdepositional igneous event, separate from the lava flows that were obviously contemporaneous with sedimentation. Although the strata were soon reassigned to a Late Triassic–Early Jurassic age, the concept that the dikes were significantly younger than the sedimentation and the lava flows has recently been challenged by the suggestion that the dikes were contemporaneous with (and indeed may have been the source of) the lava flows (Philpotts and Martello 1986).

In 1968, de Boer (1968*a*) also reported on one of his student's gravity-modeling studies that suggested the existence of a series of interior border faults preserved in stair-step fashion within the basin—faults that were now buried except for the easternmost one, which remains exposed (figure 4-4). A similar structure had been postulated to exist in Massachusetts by Willard (1952) and by Brophy et al. (1967), based on surface data.

Wessel, Sanders, and Klein: Applications of "Modern" Sedimentology

Wessel

A number of detailed sedimentological studies of modern alluvial fans were undertaken about midcentury, and the models derived from these studies were applied to the Mount Toby Formation and the Turners Falls Sandstone in the Deerfield basin in Massachusetts by Wessel, Hand, and Hayes (1967) and by Wessel (1969). With these papers, the importance and potential of sedimentary structures finally began to be realized. The studies used crossbed azimuths, clast-size trends, and clast orientation (long axes) to delineate and map three distinct fan complexes with their apices against the eastern border fault. They also documented radial dispersal/paleoflow patterns and possible internal drainage within the basin. In addition, such sedimentary structures as antidunes, parting lineation, and ripple drift were used to interpret flow directions and energy conditions during deposition.

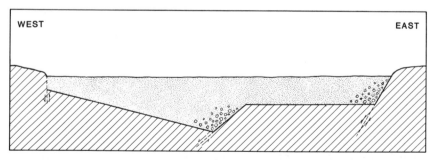

Figure 4-4. Step-faulted basin configuration of the Hartford basin during the Holyoke volcanic event, inferred on the basis of gravity modeling across the basin. (After de Boer 1968*a*, Figure 1, p. C-5. Reproduced with the permission of the Connecticut State Geological and Natural History Survey)

The fabrics and assemblages of sedimentary structures within the deposits were used to differentiate the proximal, mid, and distal fan zones. The fine-grained deposits also received attention, as Wessel interpreted the sediments in which Emerson's evaporite mineral crystal casts were found as playa-lake deposits. More permanent lakes were inferred elsewhere on the basis of the fish-bearing mudstones.

Wessel was part of a group (Hand, Wessel, and Hayes 1969) that described and interpreted one of the few convincing examples of preserved antidunes in the sedimentary record. These occur in the Mount Toby Formation, and the interpretation of them as high-discharge sheetwash deposits on an alluvial fan was considerably aided by the controlled-flume experiments (such as Middleton 1965) performed on sedimentary structures during the 1960s.

Wessel did not make use of Walker's and Van Houten's early 1960s to early 1970s papers, which were finally beginning to make sense out of the problem of the origin of redbeds. Although these papers did not deal directly with the Hartford group, they essentially substantiated and furthered many of Barrell's ideas. More importantly, both authors' ideas were well supported by the new knowledge of diagenetic processes and thus were accepted by the geological community.

Both Walker and Van Houten recognized that long-term postdepositional alterations in an oxidizing, nonmarine-to-paralic environment were necessary for the formation of redbeds; consequently, they took care to document the production of iron oxides from the destruction of iron-silicate sand grains as one major process in the formation of red-colored deposits and the dehydration of iron oxide in the clay fraction as the other. No longer was there a need to postulate a red-soil sediment source, so the barriers were removed from Barrell's arid Triassic climate and from applying the modern analogues of the alluvial fans of the southwestern United States.

Wessel's evaporites and redbeds suggested semiarid conditions of deposition, but he also noted hematitic clays and weathered rock fragments that indicated to him "intense weathering under humid conditions" (Wessel 1969, p. 142). Van Houten had published a definitive paper on the dehydration of iron hydroxides in depositional settings in 1968, but Wessel's work had probably been completed by then. Using the older ideas, Wessel (1969) concluded that the site of deposition was semiarid, but that the "moderately high, rugged source area . . . was subjected to humid weathering" (p. 146).

Sanders

Sanders's early papers on the Hartford basin were concerned exclusively with its structural aspects, but his later papers undertook detailed sedimentological interpretations. Sanders's sedimentological papers mark the beginning of a prolonged focus on the poorly exposed, fine-grained sediments in the basin. This was in large measure due to the success of papers written by Van Houten (1962, 1964, 1965) on the lacustrine Lockatong Formation in the Newark basin of New Jersey. These papers exemplified the emerging recognition among geologists of the repetitive nature of some sedimentary patterns. Van Houten had documented "cyclic" deposition in inferred alkaline lakes, attributing the patterns to variations in inferred water depths that changed with climatic trends. Olsen (1985*b*) has recently proposed that such lacustrine depositional cycles be called "Van Houten cycles."

In a SEPM (Society of Economic Paleontologists and Mineralogists) volume entitled *Primary Sedimentay Structures and Their Hydrodynamic Interpretation,* Sanders (1965) published a paper on the general characteristics and the mode of origin of turbidity current deposits. Hydrodynamics was then a new and rapidly expanding field in geology. Sanders's paper also came close on the heels of the recognition of turbidite deposits in ancient rocks (Kuenen and Migliorini 1950; Bouma 1962). Sanders followed his 1965 paper with an article (1968) that applied turbidite and deep-water concepts to fine-grained, laterally extensive deposits within the Hartford basin, proving their lacustrine origin on sedimentological grounds for the first time.

Previously, it had been recognized that these fine-grained, unchanneled, laterally extensive, and often carbonaceous deposits were not fluvial, but they had been ascribed variously to swamp, marsh, and/or lacustrine environments without substantial evidence other than the presence of fish fossils. Sanders began to recognize graded bedding and soft-sediment slumping on a centimeter scale in many of these deposits, as well as evidence such as basalt pillows and symmetric (wave-oscillation) versus asymmetric (current) ripples. He pointed out that these features did not occur exclusively in the carbonaceous and fish-bearing strata; they could be found in other sediments as well. On this basis, he began to differentiate the deposits of swamps from those of shallow lakes and from those of deep, long-lived lakes. In addition, Sanders began to appreciate the enormous size of some of these lakes, despite the relatively small proportion (probably less than 5 percent) of the sedimentary column that their deposits occupy.

In some respects, Sanders's ideas have been superseded. Because he believed in the primary red coloration of the redbeds, he thought that his maroon lacustrine deposits were of shallow-water origin and were not subject to anoxic, deep-water reducing conditions. This environment has been supported by Hubert, Reed, and Carey (1976), but the cause for the color has been reinterpreted as the oxidation and removal of carbonaceous debris during episodes of drying up of these shallow lakes. The sedimentologists who succeeded Sanders used and expanded many of his interpretations.

Although fine-grained deposits underwent close scrutiny at this time, the associated coarse-grained deposits were generally felt to be reasonably well explained by the general terms *alluvial fan* and *fluvial*. It was not yet appreciated how much room for improvement remained in these fields.

Klein

G. de Vries Klein received his Ph.D. from Yale under John Sanders in 1960. His studies complemented much of Sanders's sedimentologic work, but he subsequently took issue with Sanders's broad-terrane ideas. Having studied the Triassic–Jurassic rift-basin deposits of Nova Scotia and Connecticut, he expanded the study in 1969 to a documentation of the isolated nature of the contemporaneous rift basins—a documentation based in large part on crossbedding that suggested paleoflow into the basins from both edges. His and Wessel's were among the first studies to interpret crossbed azimuth/paleocurrent data from the Hartford group. Others, including Emerson (1898*b*) and Lehman (1959), had made local observations on crossbedding and paleocurrents but had not used them as a basis for significant interpretations.

Klein (1968) also published a field-trip guidebook article that covered deposits of the East Berlin Formation, in which he interpreted black and gray mudstones as lacustrine deposits and recognized that they alternated cyclically with red fluvial mud-flat mudstones. The potential of this idea was never fully crystallized in Klein's papers. While he made good use of sequences of lithologies and facies, he never collaborated with Sanders, who had considered the environmental significance of sedimentary structures but had not addressed facies sequences in the rocks. These two aspects of the sedimentology of the Hartford group's lacustrine sequences were not melded until 1976, when Hubert, Reed, and Carey picked up where Klein and Sanders left off.

Advances in Other Areas

Tectonics

The 1970s saw increasing acceptance of the tenets of plate tectonics, and thus increasingly sophisticated tectonic and structural interpretations were applied to the rift basins, including the Hartford basin. Dewey and Bird (1970) began to correlate the maximum age of the Atlantic oceanic crust, the age of initial continental rifting, and the formation of the extensional rift basins, and to view the entire assemblage as an integrated process of continental breakup.

Slightly later, it began to be recognized that plate tectonics could not account for all aspects of the observed structure, so the influence of previous structures on the orientation and character of the rift basins and their faults and dikes was investigated. Wise (1982) summarized and advanced the idea that the Hartford basin might owe its main north–south orientation to previously existing structures associated with earlier Appalachian tectonics. The N30°E faults that offset the eastern border fault in places and control the emplacement of the basaltic dikes reflect the regional extensional stresses associated with continental rifting more closely than does the general axis of the basin.

This parallelism between the rift basins of eastern North America and the structural trends of the Appalachians had been noted and puzzled over by various geologists since the mid- to late 1800s, when Dana had theorized that the rift basins were compressional structures associated with the formation of the Appalachians. Now it was becoming apparent that the relationship was not one of different responses to the same cause, but rather one in which rift-basin trends had been inherited from the structural fabric of the previous orogeny.

Correlations Across the Atlantic

Following the acceptance of the ideas of continental rifting and global tectonics, geologists began to compare rift basins and their deposits on opposite sides of the Atlantic. Two summary papers on such comparisons (Van Houten 1977; and Manspeizer, Puffer, and Couzminer 1978) evolved out of a project put together by W. H. Kanes to study the geology of Morocco. While not always agreeing on specific points, these studies pointed out both similarities and differences in the structural and sedimentological responses to rifting that occurred in different areas. The Moroccan rift basins also demonstrate an inherited control on basin evolution. The two principal Moroccan basins were filled primarily from highlands located at the ends of the basins, rather than from source areas along the sides, but many of the sedimentary environments were similar to the trans-Atlantic basins.

Palynology: Age and Depositional Environment

Although early attempts to find microfossils and spores in Newark Super-group sediments had produced "largely negative results" (Sanders 1968), B. Cornet found a rich palynomorph assemblage in certain strata of the Hartford basin in 1970 (Cornet, Traverse, and McDonald 1973; Cornet and Traverse 1975). The reports of these findings suggest that the paleoflora—at least from the lacustrine and closely associated deposits—was dominated by conifers.

More importantly, they favored reassignment of the sediments to the latest Triassic and Early Jurassic time periods, a determination that had been discounted and often forgotten during the preceding three quarters of a century.

These studies suggested that the Triassic–Jurassic boundary, marked tentatively by a major shift in the palynomorph assemblage, could be placed just below the Shuttle Meadow Formation. They also suggested that an appreciable erosional unconformity lay within the basin fill at this boundary, between the Talcott Basalt and the Shuttle Meadow Formation (figure 4-5). The physical evidence for this unconformity consists primarily of the localization of basalt flows of the Talcott episode to Sanders's Gaillard graben, which Cornet took to indicate the syndepositional subsidence of the graben and the preservation of flows within it while they were eroded outside the graben. It could equally well be that the flows were only deposited within the

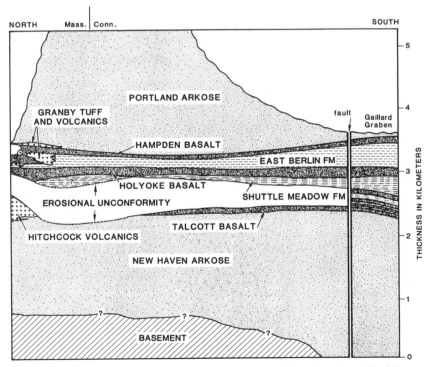

Figure 4-5. North–south schematic restored cross section along the axis of the Hartford basin, showing an inferred erosional unconformity within the Hartford group. (After Cornet and Traverse 1975, Figure 3, p. 3. Reproduced with the permission of *Geoscience and Man*)

subsiding graben; erosional relief superimposed on the flows has not been documented, except locally (Gray 1982).

Colbert and Gregory (1957, p. 1460) had noted a faunal break at the level of the lava flows in both Connecticut and New Jersey that may coincide with this unconformity and/or with the Triassic–Jurassic boundary. According to these authors, the remains of quadrupedal phytosaurs and pseudosuchians are found below the lavas, whereas the vertebrate fauna above these beds consists primarily of bipedal dinosaurs.

From their data, Cornet and co-workers were able to draw two conclusions regarding depositional environment. The first followed the 1973 study that had included observations of a rich fish fauna associated with "varved organic micrite" and pyritic lake sediments; the conclusion was that there had been meromictic (stratified) lakes in seasonal climates. The second conclusion demonstrates the long-lasting effects of Krynine's work, in that Cornet and Traverse (1975) made their conclusions fit into his humid, seasonal savanna-type climate. The climatic implications of their pollen assemblage apparently could have been interpreted in several ways, including seeing the assemblage as consistent with arid to semiarid conditions, but Cornet and Traverse (1975, p. 1) opted for a "warm, seasonally wet and dry climate." In retrospect, their data would probably fit equally well with the subsequent climatic models of Hubert and his co-workers, who finally found sedimentological evidence to support many of Barrell's concepts of semiaridity for the formation as a whole. Hubert, Reed, and Carey (1976) have, however, chosen to use the palynological interpretation of a "wet" climate as corroboration for the hypothetical origin of the lacustrine cycles during episodes of higher precipitation within the overall semiarid climate.

Fluvial and Lacustrine Sedimentation

From 1976 through the early 1980s, Hubert and his co-workers and students published a number of papers on various aspects of the Hartford group. These papers looked at assemblages of primary sedimentary structures, repetitive sedimentation patterns, and paleocurrent indicators in order to reconstruct detailed depositional environments and paleogeographic distributions of those environments.

Lacustrine Sedimentation

The first paper (Hubert, Reed, and Carey 1976) dealt with the floodplain and lacustrine deposits of the East Berlin Formation. The researchers, aided by continued highway projects that provided road cuts through these fine-grained and usually poorly exposed strata, confirmed many of Sanders's earlier interpretations of the rocks, while adding significantly more detail to the paleogeographic maps and sedimentary history of the deposits.

Two types of lacustrine deposits were identified. As interpreted by Hubert's group, the first type of deposit was produced in small, shallow lakes on the floodplains. Periodic dryness allowed oxidation of accumulated plant debris and thus the eventual formation of red coloration. Preferred orientations of current ripple marks were inferred to reflect the dominant wind direction, locally refracted against irregular coastlines.

The second type of deposit was cyclic and was interpreted as recording the initiation, expansion, and contraction of basin-wide lakes (figure 4-6). The central part of the cycle, dolomitic mudstone containing pyrite and well-preserved fish fossils, was taken to represent anaerobic deposition below the thermocline in deep oligomictic or meromictic (usually or always stratified) conditions, when the lake was at its full extent and relatively deep. Near-shore or beach deposits were inferred for the parallel-bedded sandy units near the top and base of the cycles. The millimeter-scale couplets in the mudstones were interpreted to indicate chemical precipitation in alternating dry and wet climatic cycles—possibly seasonal and, therefore, possibly "varves." It was suggested that lacustrine basin-floor slump deposits recorded earth-quake shocks generated by syndepositional movement along the eastern border fault.

On a larger scale of interpretation, Hubert, Reed, and Carey suggested that the lacustrine cycles were initiated by long-term climatic fluctuations from tropical, semiarid climates to tropical, humid climates associated with significant increases in precipitation. It is puzzling why this climatic change should have coincided with changes in volcanic activity (basalt outpourings) and possibly with changes in tectonic activity (suggested by fewer border conglomerates associated with the lacustrine sediments), but to date, evidence to support alternative explanations is largely lacking.

Hubert, Reed, and Carey may have been following the lead of Van Houten (1962, 1964), who had assembled strong evidence, in eighty measured cycles, for cyclic variation of a slightly different pattern within the older Lockatong Formation (Newark basin, New Jersey) in a more tectonically stable setting. Van Houten's paper attributed the Lockatong cycles to climatic changes that varied as a function of the precession of the earth's spin axis. He noted, however (1962, p. 572), that "prejudice in favor of a cyclic climatic control, rather than periodic downwarping, stems largely from an assumption that variations in climate are apt to be more regular and persistent than intermittent subsidence." Although the Lockatong cycles may have been climatically controlled, not all lacustrine cycles need be—especially where there is evidence for contemporaneous tectonic activity.

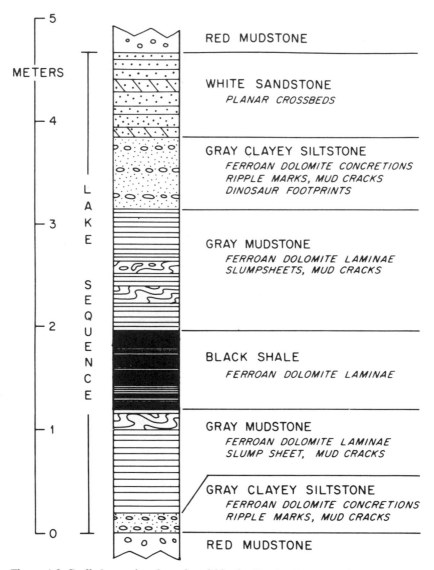

Figure 4-6. Cyclic lacustrine deposits within the East Berlin Formation, interpreted to record the initiation, expansion, and (finally) regression of a lake environment. (From Hubert, Reed, and Carey 1976, Figure 7. Reprinted with the permission of the *American Journal of Science*)

The suite of authigenic minerals in the lacustrine deposits of the Hartford basin suggested to Hubert, Reed, and Carey that these lakes were alkaline, although not to the same extent as those that formed the analcime-rich Lockatong deposits in the Newark basin. Traces of these evaporites are not uncommon in Connecticut, but they are not present in large volume.

An important point to note here is that most of these lacustrine deposits, both playa and perennial, were interpreted at this stage primarily on the basis of models of vertical sequences that had been derived directly from the ancient record, rather than by analogy to specific modern environments. Many aspects of these models continue to be compelling, as studies of modern lakes provide analogies to the ancient strata and as more data are extracted from the rocks. Part of the problem (and the reason why the models were derived directly from the ancient record) is the fact that modern lacustrine and playa environments have been studied primarily from geomorphic and geochemical angles; until recently, little was known about their assemblages and about the patterns of their sedimentary structures.

Parts of the East Berlin Formation have since been interpreted as a "sandflat–mudflat–lake complex in an enclosed basin," using a predominantly playa depositional model (Gierlowski-Kordesch and Demicco 1983; Demicco and Gierlowski-Kordesch 1986). It would appear that the sandy units with climbing ripples, flazer, and wavy bedding that Hubert, Carey, and Reed ascribed to shallow lake deposits are the ones being interpreted by Demicco and Gierlowski-Kordesch as products of distal alluvial-fan and playa-margin sand-flat and mud-flat environments. It is not clear whether (and how) the other lithofacies interpretations of the two studies are compatible, as the "typical" measured sections through the lacustrine deposits (compare figures 4-6 and 8-14) are significantly different as portrayed in each of the two studies.

Hubert, Gilchrist, and Reed (1982) have since reinterpreted some of their 1978 conclusions, suggesting that some beds formerly assigned to fluvial and floodplain environments are in fact playa and sheetflood deposits; however, it is not always apparent which beds are being reinterpreted or what evidence is being used for the reinterpretation. The impetus for these new interpretations has been the recently published study by Hardie, Smoot, and Eugster (1978) of modern closed-basin playa environments, which has provided a new environmental model for interpreting such sedimentary deposits.

Fluvial Sedimentation

Hubert, Reed, and Carey (1976) also addressed the fluvial (channel and floodplain) deposits associated with these lacustrine deposits. Allen (1965) and Visher (1965), among others, had only recently begun to show that untapped data existed in the detail of sedimentary structures and vertical sequences found in fluvial deposits; however, the first conference on fluvial sedimentology (1977) was still a year away. (See Miall 1978 for a history of fluvial sedimentology.) As a result, the fluvial channel deposits received scant attention. Braided-stream deposits were characterized only by their coarse nature and by the predominance of parallel bedding in them, whereas meandering stream interpretations were suggested only by the evidence of fining-upward sandy sequences and of scattered to radiating crossbed paleoflow indicators, with large variances in individual outcrops.

The floodplain mudstones, in contrast, were given somewhat more detailed treatment. In addition to the indicators for subaerial exposure that had been used in interpretations for a century and a half (mud cracks, vertebrate tracks, and raindrop impressions), Hubert and his co-workers began to recognize carbonate paleosols, claystone rip-ups, and thin, graded beds, all deposited from suspension during inundations of the floodplain.

Also described were floodplain slumps, all of which are unidirectional to the southwest. Although their characteristics and mode of origin were not elucidated, the slumps coincide with regional crossbedding vectors in indicating a southwesterly paleoslope.

The concepts of fluvial sedimentology advanced in two years, and more outcrops of this facies were studied in detail. In 1978 Hubert and his co-workers put out a SEPM guidebook covering parts of the Hartford basin. Complex, nonsequential patterns of sedimentary structures were described, suggesting braided conditions in rivers that deposited the New Haven Arkose in the main basin.

Correlative deposits in the Pomperaug outlier display fining-upward sequences, and thus a facies change from alluvial fans to braided streams to meandering streams may have occurred in a southwestward direction from the border fault (figure 4-7). Clast-size trends were shown to be continuous through the outlier, supporting the idea that it was continuous with the main basin during deposition.

Hubert et al. also presented the first detailed set of basin-wide paleocurrent measurements (for the New Haven Arkose, see Hubert et al. 1978, p. 42, Figure 19) and reconstructed the regional paleoslope and paleogeography. The complexity of the paleocurrent vectors in the southern parts of the basin is interpreted as "back-and-forth shifting of the braided-river belts in response to alluviation on the gradually subsiding valley floor" (Hubert et al. 1978, p. 41). Another possibility is topographic control by syndepositional faulting within the basin—internal structures such as the Gaillard graben.

More recently, Little (1982) has reported occurrences of armored mud

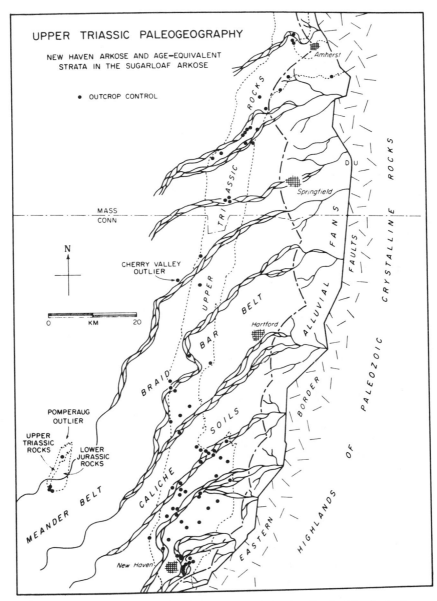

UPPER TRIASSIC PALEOGEOGRAPHY

NEW HAVEN ARKOSE AND AGE–EQUIVALENT
STRATA IN THE SUGARLOAF ARKOSE

• OUTCROP CONTROL

Figure 4-7. Reconstruction of the paleogeography in the Hartford basin during stages of alluvial-fan and fluvial deposition. (From Hubert et al. 1978, Figure 20. Reprinted with the permission of the Dept. of Geology and Geography of the University of Massachusetts)

balls from the Turners Falls Sandstone in Massachusetts. These are common in modern environments, but relatively rare in the ancient record, especially in fluvial deposits. Little suggested that they formed in a distal fan environment as steam channels cut laterally into lake and floodplain mudstones, creating soft, plastic, mud interclasts that quickly became armored as they were transported along the sandy riverbeds.

Paleosols and Climate

The possibility that carbonate paleosols—often attaining the end-member form of carbonate-plugged horizons or caliches, and most common in the New Haven Arkose—were present in the Pomperaug outlier was suggested by Schutz (1956) in a senior thesis paper at Yale. However, modern and recent examples of carbonate paleosols were described and models of their processes and genesis were derived only beginning in the mid-1960s, and they were only fully recognized and described in the Hartford basin by Hubert (1977, 1978). Although caliches are relatively common in the Hartford group, their presence in and significance to the Newark deposits had been essentially overlooked by previous geologists. This may be an example of the principle that geologists often fail to see what they are not looking for in the field, although it probably also illustrates the phenomenon that geologists commonly do not describe what they cannot confidently interpret.

The paleosols provided evidence for Hubert's climatic interpretations: semiarid with long dry and short wet seasons. This climatic regime is essentially the opposite of Krynine's regime (humid with long wet and short dry seasons), and it was envisioned as having undergone temporary modifications from time to time that would allow for the development of the large lakes.

The proposed climate was supported by work on the genesis of color in the Hartford group (Hubert and Reed 1978). This work applied the ideas of Walker and Van Houten to petrographic and geochemical observations, and it demonstrated the diagenetic processes that created red hematite from the following sources:

1. The dehydration of detrital limonite
2. The iron derived from the dissolution of iron-silicate grains
3. The oxidation of common ilmenite and magnetite
4. The iron released by the replacement of iron-bearing grains by dolomite cement

Fan and Sheetflood Deposits

One of the more recent papers of this series appeared in 1980 (Stevens and Hubert 1980) and addressed the earliest deposits in the northern

Deerfield basin. Clast-size trends and paleocurrent indicators in the Sugarloaf Arkose (below the basalt flow) pinpointed fan apex locations that were
similar to those defined by Wessel (1969) in the overlying Mount Toby
Formation. However, because the basalt and younger sediments cover proximal Sugarloaf deposits, the exposed Sugarloaf represents distal-fan, fluvial,
and floodplain environments. A fourth fan apex was located, occurring
against the northern margin of the basin, with the fan spreading southward.
Lacustrine deposits were also recognized in the northern part of this basin,
occurring at the top of the Sugarloaf sequence, just below pillow basalts.

Stevens and Hubert showed that paleocurrent and clast-size trends indicate a former extension of the sedimentary deposits westward beyond the
present boundary of the basin. Although they used the term *broad terrane*,
they did not specifically advocate continuity of the deposits with those of the
Newark basin, and they demonstrated longitudinal drainage southward
down the axis of the basin in the presently preserved deposits. Weddle and
Hubert (1983) later presented detailed petrologic data indicating synsedimentary
separation of the Newark and Hartford basins.

Another type of data, new to the Hartford basin, was the recognition of
Scoyenia burrows. These trace fossils, apparently made by animals related to
crayfish, were used as indicators of a floodplain environment.

Concurrent with his work in the Hartford basin, Hubert has carried on an
active program of research in the Triassic–Jurassic rift-basin deposits of the
maritime provinces of Canada. In 1982, Hubert and Hyde published a paper
documenting playa and sheetflood deposits in the Fundy basin of Nova Scotia.
The modern process of sheetflooding that occurs on sand flats near the
transition from alluvial-fan to playa environments has been known since the
late 1800s (Hogg 1982), but their deposits have been little studied. Thus, the
model of sheetflood deposits on sand flats is another that had been derived
in large part from interpretations of the ancient record. The previous lack of
this depositional model was a factor in the decision of Byrnes and Horne (1974)
to interpret parts of the Hartford group as marine—specifically the deposits
that were internally horizontal-parallel bedded, often graded, and of uniform thickness laterally for outcrop-scale distances. These beds are now
inferred to be sheetflood or sheetflow deposits (Hubert, Gilchrist, and Reed
1982; Demicco and Gierlowski-Kordesch 1986; LeTourneau 1985a, 1985c).

Recent and Ongoing Work

Other than the expansion of the application of the playa/playa-lake environmental model, recently published work has done little to alter the
conclusions of Hubert and his co-workers. Olsen, McCune, and Thomson
(1982) put together ideas on stratigraphic correlation of the Newark Super-

group between the basins along the entire Atlantic seaboard. Although Olsen (1983) has since suggested that their precise basin-to-basin correlations may have been premature and incorrect, the multidisciplinary approach, underpinned by the researchers' own work on the fish fossil correlations, offers support for the idea that the age of the sediments within the North American rift basins becomes younger in an erratic but significant fashion from south to north. Nadon and Middleton (1984) have since suggested that palynological age dates indicate that the northward sequence is broken and repeated beginning at the Gulf of Maine.

Elsewhere, Kaye (1983) published a paper on the discovery of a third small infaulted outlier of red arkoses, siltstones, and conglomerates. This outlier, named the Middleton basin, is located 17 kilometers north of Boston and considerably east of the main Hartford basin. Its main significance lies in its implications for the breadth of the mesh of isolated basins along the rift trend.

Another short, recent study was that of Parnell (1983), who convincingly documented pseudomorphs of skeletal halite (long noted but never rigorously documented or explained) in lacustrine mudstones in Massachusetts. Parnell suggested that these crystals grew subaqueously and displacively within the sediment, without requiring complete evaporation of the lakes for formation.

A number of studies, published as yet only in abstract form, will probably have been completed by the time this volume is published. Most of them have been undertaken in response to renewed interest in the Newark basins as potential sources of oil and gas—first by the oil and gas industry (for economic reasons), second by the U.S. Geological Survey (which is charged with knowing about existing or potential resources), and third by academia.

A measure of the economic interest is the number of abstracts published recently that address the problems of various sediments' organic content and level of maturation (critical to the generation of economic hydrocarbons) in the Newark Supergroup. Among these are the studies of High (1985), Kotra et al. (1985), and Schamel and Hubbard (1985). Results to date are ambiguous, with the organic content of the known potential source beds being marginal and the trends of maturation erratic. More recently, Pratt, Vuletich, and Burruss (1986) have suggested that good, but very thin, source rocks may exist within the Hartford basin.

Concurrent with the studies that are directly related to hydrocarbons, advances are being made in paleontology and sedimentology. Two extended studies specifically related to sedimentology have great potential. LeTourneau (1984, 1985a, 1985b) has been deciphering the deposits of the Portland Arkose and has distinguished "wet" and "dry" facies/environment assemblages, and episodes of rejuvenation of fan sedimentation. The other program is being undertaken by the U.S. Geological Survey and includes a multifaceted sedimentology study program. Recent abstracts (Smoot et al. 1985; Smoot and Olsen 1985) have described different types of shoreline deposits (a facies building block that has been missing or ill-defined in the lacustrine models to date) and different types of mudstone deposits.

Summary

The Course of Progress

The foregoing four chapters cover the last 175 years, years that included periods of gyrations and episodes of stagnation in geological thought as applied to a single sedimentary basin-fill sequence. Throughout this history, geologists' interpretations have been limited not so much by the geologists' individual capabilities as by the degree of ignorance that they were working in. Part I of this book emphasizes the evolution of geological concepts, but these are usually difficult to dissociate from the individual geologists who espoused them—especially during the early years, when there were few geologists and widely divergent ideas.

Scientific advances usually occurred slowly, although they grew more frequent after the first century covered here. Small pieces of data or minor changes in interpretations were added incrementally and often inconspicuously, until someone with a broad enough perspective could detach his nose from the outcrop, step back, gather the loose ends of seemingly unrelated ideas, and put them together with newly evolved general concepts into a quantum jump of progress. Unfortunately, as often as not, this person was either wrong or too early with the ideas and was relegated to obscurity by the conventional wisdom of the time. Conversely, geologists who would eventually be proved wrong often enjoyed temporary but tremendous success. As noted by Greene (1982, p. 292), "geology was a larger and more difficult enterprise than had previously been imagined." This statement could easily be expressed in the present tense.

"Sheer proof" is rare in geology: there is almost always more than one explanation for a given phenomenon. Proof of a hypothesis commonly rests on a few critical pieces of data, supported by the weight of all the circumstantial evidence an author can think of. Words such as *suggest* and *imply* are more common than might be desirable, but at this stage in the progress of the science, the basis on which any models are built—even our computer models—is still subject to revision.

On the other hand, each theory or interpretation has served a purpose, whether it eventually proved to be right or wrong. Each became a focal point for more study, either as a base on which new ideas could be built or as a target that new ideas and data would be generated to attack (and often as a combination of both). As T. H. Huxley put it, "the attainment of scientific truth has been effected, to a great extent, by the help of scientific errors" (Huxley 1887, p. 57). Each theory became part of the broad (though often invisible) base of scientific knowledge on which we all stand.

In the past, when there were few unambiguous data to support definitive interpretations and few models to serve as standards for making definitive comparisons, speculation played an important role in the advancement of the science of geology. Tenuously supported hypotheses served to focus study, as both unconscious and deliberate attempts were made to prove or disprove the hypotheses. Such ideas might serve a similar function today, although we have more rigorous means of developing and defending theories. Unsupported musings often seem out of place in a paper containing otherwise well-documented interpretations, however, and they are usually excised by reviewers. While there is clearly no usefulness in unwarranted speculation, experience with a subject sometimes gives the investigator an insight into an aspect of the work that cannot be immediately followed up on but that may yet help to point the way toward future productive research. While the distinction between the well supported and the speculative should be made clear by an author, it should be the task of the reader to separate them and to use each appropriately. An author's imagination should not be totally stifled.

The general approaches to techniques of sedimentological study have changed in conjunction with the changing concepts. Middleton (1978) has divided the history of sedimentology into five periods. The new techniques used by each period are listed as follows:

1. The acceptance of actualism (to 1830)
2. The acceptance and use of facies concepts and petrologic techniques (1830–1894)
3. Grain-size analyses and laboratory experimentation (flume studies) (1894–1931)
4. Basin analyses and facies use (1931–1950)
5. The current use of sedimentary structures and studies of modern environments (1950–present)

Studies of the Hartford group generally parallel this sequence, although the timing in specific instances is somewhat different.

Gaps in Present Knowledge

Regarding the Hartford basin, there is still much good geology to be done, especially in the realm of sedimentology and basin analysis. Three unsolved problems that may even now be under study come to mind readily: the detailed sedimentology of the different fluvial regimes present; the significance and genesis of the isolated evaporite crystals that apparently formed within the perennial, theoretically deep-water lacustrine sequences; and the curious fact that paleoslopes (as indicated by the majority of the published measurements of lava paleoflow) are downhill to the northeast—the reverse of the documented fluvial paleoflow (figure 4-7)—toward the border-fault alluvial fans. At least seven different studies mention observations of this last phenomenon, but no one has yet made integrated paleogeographic and tectonic inferences from basalt paleoflow directions. Such a study would be especially useful now that Philpotts and Martello (1986) have suggested the locations of the feeder dikes for each of the three flows.

Possibly related to the third problem named above is another area that is ripe for research: the climatic, tectonic, and/or volcanic factors that caused the initiation of perennial lacustrine conditions. These have yet to be sorted out and documented. Climatic causes—especially if these are based on worldwide rotational precession cycles—should be recorded in correlative sedimentary responses between the basins. Tectonic events may also be correlative, but probably not beyond neighboring basins. Olsen has apparently made comparisons between cycles of the lacustrine strata in the Hartford and Newark basins (see Olsen 1986), but the data and conclusions have not yet been published.

Hitchcock (1843) suggested that the absence of growth rings on one specimen of fossil wood indicated either rapid growth or a warm climate without significant seasonal variation. Although small-scale laminations in the lacustrine sediments of the Hartford group have been tentatively linked to short, possibly yearly, climatic variations, this evidence has not been compared with that from the characteristics of the rare fossil wood (which itself has not been addressed for some time).

Observations from different basins may have some bearing on the question of the origin of the lakes. Lindholm (1978) suggested that the localization of lava flows against the border fault of the Culpeper basin (obviously not climatically controlled) was caused by internal tilting of the basin floor and a consequent rearrangement of the paleoslope. Significantly, the lacustrine conditions in this basin occurred in the same area and may have originated through the same controls of asymmetric, irregular subsidence of the basin. The possible continuous extension of lacustrine sediments and basalt flows across the border fault suggests that a period of tectonic stability followed the rearrangement of the paleoslope.

Regional paleoslope within the Newark basin was toward the northwestern border fault during most of its history (Van Houten 1965); the magnitude of faulting was sufficient to create alluvial-fan paleoslopes away from the fault only locally. Recently, LeTourneau (1985a, 1985c) has suggested a similar geometry for the Hartford basin floor during some of the lacustrine sequences within the Portland Arkose. He has documented a thickening of the dark mudstones eastward toward the border fault and alluvial fans, inferring an easterly paleoslope at the time of deposition of this facies.

The internal paleoslopes in these rift basins were subject to a delicate balance, and were easily reversed by irregularities in the tectonic/subsidence framework. Such reversals might show up in the disruption of established drainage—that is, in the creation of lakes. Ratcliffe (1971) has suggested a hinged mechanism (wherein one end of a fault has greater motion than the other) for the Ramapo fault bordering the Newark basin. If a similar mechanism operated along the border fault(s) in Connecticut and Massachusetts, with different ends or portions undergoing different displacements at different times during deposition, reversed longitudinal drainage of the basin may have been possible at times.

Bain suggested that evidence exists for such hinged fault motion locally along the northern parts of the border fault in Massachusetts (Bain 1957a, 1957b). While the measurements of southwesterly lake-bottom paleoslopes (Hubert, Reed, and Carey 1976) might argue against this scheme, the paleoflow measurements of the basalts, as well as LeTourneau's measurements of eastward-thickening black shales, might support it. Hubert and his co-workers may in fact have been measuring slumps only on the steeper, near-shore depositional break in slope, rather than from the actual basin-floor paleoslope. This type of data may bear on possible tectonic models for the origin and cyclicity of lacustrine sedimentation.

The intragroup unconformity postulated by Cornet and Traverse (1975) has yet to receive much attention. Is it real? As cited in Robinson and Luttrell (1985), Cornet has also suggested on the basis of palynological data that there is a correlative and significant hiatus in the Deerfield basin. This is placed following the deposition of thick lacustrine units, and preceding the tectonism that produced a "slump zone unconformity" within the partially lithified lake strata, as the coarse Mount Toby Formation was introduced into the basin. Although Gray (1982) has suggested that Mesozoic erosion is responsible for the absence of the Talcott Basalt in the basin from East Granby, Connecticut, north, no sedimentological evidence for this unconformity has yet been published. Such an unconformity, with or without an attendant reversal of drainage that may have caused some of the lacustrine conditions in the rifts, could be related to regional reversals of paleoslopes during the uplift of the new continental margins immediately after final separation of rifting continents. No one has yet studied the effects of such tectonic events on rift-basin sedimentology.

On a larger scale, migrating flexural "bulges," such as those postulated by Watts (1981) for post-rift continental margins, have not been studied extensively in the post-Triassic structural record. The record of these features would be erosional and structural rather than depositional; thus, the "Palisades disturbance" that caused postdepositional deformation and erosion of the Hartford group may be related to such mechanisms, as suggested by Austin et al. (1980). This hypothesis is especially attractive in the absence of other orogenic events on the passive continental margin.

No definitive age dates have yet been derived from the New Haven Arkose. Perhaps if significant hydrocarbons are found in the basin, the resulting subsurface data will provide unweathered rock (core or cuttings) for palynological age dating, as well as allowing for the compilation of comprehensive formation isopach maps. A good first step toward accomplishing the latter task could be taken for the formations between the basalt flows by using surface data. Age dates and isopachs would help in deciphering possible interformational unconformities, especially if significant erosion occurred.

Subsurface data would also be useful in reconstructing facies distributions, which—along with the isopachs themselves—should help answer the question of syndepositional faulting within the basin (as in the case of the Gaillard graben). Facies thickness and distribution trends should begin to illuminate the record of border-fault/tectonic activity as it varied along the length of the basin and through time. LeTourneau (1984) and LeTourneau and McDonald (1985) have inferred at least two cycles of fan rejuvenation (and thus, possibly, of fault reactivation) recorded in the Portland Arkose; in certain places, alluvial fans were contemporaneous with deposition of the lacustrine deposits, whereas elsewhere the lake deposits abut the border fault (Krynine 1950). The patterns and significance of these have yet to be sorted out for the entire basin.

The differences in the different types of mudstones are just beginning to be examined (Smoot and Olsen 1985). This is a difficult study, in view of the poor exposures of the mudstones, but it should be rewarding. Again, the study is hampered by the paucity of modern geological models for comparison. Although processes of wetting and drying that form crumb fabrics or "peds" have been known to soil scientists (Brewer 1964), these processes and products are difficult to recognize in the rock record, and geologists have made little use of them.

The significance of Emerson's Chicopee Shale and Longmeadow Sandstone, now defunct as formations but still present as distinctive lithologies within the Turners Falls Sandstone of Massachusetts, needs to be addressed. As formerly mapped, the fine-grained Chicopee Shale would have extended southward into the middle or lower zones of the Portland Arkose of Connecticut, where it may correlate with some of the lacustrine sequences of this formation. There may also be some significance to the apparent southward stratigraphic rise of the Chicopee Shale as mapped against the Holyoke basalt flow (Emerson 1917). This pattern may be present in other lithologies, such as in the limestone unit between the basalts in Connecticut (Krynine 1950); if so, this would suggest gradual longitudinal shifts in depositional environments, reflecting tectonic rather than climatic influence on deposition.

Postscript

My own interest in the Hartford group began at the age of seven or eight, when a bulldozer caked with "red clay from Cheshire," as the operator told me, was brought to our rural home in the metamorphic terrane of the western Connecticut highlands. The clay was much more plastic than the local material, and since it made significantly better mudpies, the incident made a lasting impression. This lead-in was not reinforced, however, until some fifteen years later, when I had the pleasure of studying as a graduate student under Dr. Franklyn B. Van Houten in the Triassic of Morocco. I have maintained a peripheral interest since then, despite the fact that job opportunities have led me elsewhere entirely.

It is perhaps too easy to speculate, postulate, decipher history, and propose work for the Hartford basin when writing at the kitchen table in the mountains of New Mexico, some 3,000 kilometers away. By the same token, however, the distance offers a perspective not readily attained when one is trying to locate outcrops through the rain and vegetation or trying to unravel and write up field notes in time to meet an administrative or abstract submission deadline. It is my hope that the historical and personal viewpoint adopted and given expression in Part I is of value both to students and to those currently working in the basin.

PART II

SEDIMENTOLOGY OF TRIASSIC–JURASSIC RIFT-BASIN DEPOSITS

Introduction

Scope and Purpose

Part II examines the sedimentology of the seven generalized depositional environments found in the Triassic–Jurassic mesh of rift basins located along the continental margins adjacent to the North Atlantic Ocean.

The examinations are limited to onshore, exposed basins, where outcrops have allowed detailed sedimentological field studies. Numerous other rift basins or grabens are present in the subsurface of England, Morocco, the Atlantic coastal plain of North America, and the offshore continental shelf (Klitgord and Hutchinson 1985), but few of these are discussed.

Emphasis is placed on the techniques that have been used for study and on how geologists have applied existing models of depositional environments. The chapters are presented as case histories, and the reader is assumed to have a basic knowledge of sedimentology. Stratigraphic nomenclature and correlation are not addressed—not because they are not important, but because they are not central to the subject matter of depositional environments, facies patterns, and the formulation of sedimentological interpretations.

Because this book is not meant to be an exhaustive review of existing rift-basin literature, only the more recent, detailed sedimentological studies (or older ones which present good data that have not been recently interpreted) are introduced here. This requires some reiteration of recent sedimentological studies from the Hartford basin. Although the chapters are not intended to be purely descriptive, a summary of much literature on Triassic-Jurassic rift-basin sedimentology is necessarily presented. It should be noted, however, that the full range of characteristics of the deposits of each environment is not illustrated in these rift-basin settings, even though significant variability exists.

As was obvious in Part I, without detailed modern analogues of depositional environments to serve as models for interpretation, early geologists had no real limits on possible interpretations. As early as 1817 (p. iv), Maclure noted that, in the absence of models or a "scale by which we can measure their [hypotheses] relative merits . . . one conjecture is perhaps as good as another. . . ." As quickly as studies of modern environments were published, they were applied to the ancient record by model-starved field workers. The new models were often applied indiscriminately because of the lack of alternatives, or sometimes because they were in vogue.

Environmental models are still proliferating, but this may be a good time to make a second-generation assessment—to look at the interpreted ancient record in order to decide what parts of each environment are best preserved and what features in the preserved sediments are most diagnostic of each environment. An attempt is made here to reach behind the interpretations and to examine the primary data and logic that geologists have used as the basis for their interpretations. Some of the data are relatively inconclusive, and the finality with which the eventual interpretations have been presented in such cases is misleading. Without access to all of the basins, however, few attempts are made here to second-guess an author's data or to reinterpret conclusions, and the most recent interpretations are usually accepted as valid.

There is, of course, an obvious danger in this approach: the present interpretations, like past ones, are subject to reinterpretation, whereupon the characteristics singled out become diagnostic of a different environment. For example, some of the laterally continuous mudstone beds in the rift basins have been viewed as floodplain, playa/lacustrine, or (more recently) playa/alluvial-fan/mudflat deposits. Insofar as this book is a study of methods, however, the interpretations discussed here need not be the final truth in order to illustrate the points being made.

The seeming certainty with which most conclusions are presented brings up another point: ideas that sound definitive in print are often the end product of several sequential levels of interpretation, in the course of which extraneous data and uncertainties are filtered out. The first level of filtering occurs as the data are collected, when the investigator must decide what is worth noting or photographing. The second level takes place in the synthesis or transition from notes to manuscript. Uncertainties are rarely published and often are eliminated at this stage. The reader of a manuscript, by subjectively assessing the data and conclusions, then imposes the third interpretive filtering stage. A geologist of limited experience—or one with an axe to grind—can, at any of the three stages, significantly bias the conclusions reached. The next geologist going to the outcrop, paper in hand and ready to apply concepts, can easily become distracted by the abundant, unfiltered detail present in the rocks.

This is not to say that filtering is undesirable. In order to envision gross-scale patterns, to isolate trends, to integrate numerous outcrops, and to interpret basin-wide environments and tectonics, both filtering and synthesis must take place. Besides, there is not enough publication space for (and no one today would bother to read) excessively detailed lithologic descriptions. When reading the literature, however, a geologist should be aware of the stages of filtering that have taken place.

In studying sedimentary deposits, one can distinguish several levels of sedimentological systems. From specific to broad scale, these levels might be listed as follows:

- Event/deposit—specific flow regime (such as energy and turbulence levels)
- Subenvironment (such as debris flow on an alluvial fan)
- General environment or depositional facies (such as alluvial fan—typically but not always what is meant by the term *depositional environment*)
- Facies relationships and lateral variations of the systems (such as alluvial fan grading distally into playa)
- Changes due to the various controls on the first four levels (such as base level, tectonic, or climatic changes)
- Integrative basin analysis of all of the previous five levels (such as paleogeography through time)

Each level provides the building blocks for the succeeding one. Most of these levels are illustrated in this book by assemblages of deposits found in the rift basins, but the emphasis here will be on the intermediate levels.

Many workers, both early and recent, have made generalized environmental interpretations on a very preliminary basis—judging, for example, that poorly sorted, crudely bedded conglomerates are probably alluvial-fan deposits. While such conclusions are usually correct on a gross scale, it is often possible to milk more information from the outcrops and to be much more definitive—identifying, for example, sheetflood, debris flow, and/or sieve deposits within the alluvial-fan setting. Similarly, it is possible to differentiate among the different types of rivers that left deposits in an overall fluvial environment.

The aim here is to show that there is enough variability in modern environments (and in the sedimentary deposits attributed to them) to render a cookbook approach generally inadequate. The characteristics of an environment and its deposits can be generalized, but the current ability to recognize fine details of sedimentary structures and their associations also allows for the recognition of considerable variability within those generalizations.

Not all of the different facies in each rift basin have been studied in sedimentological detail. The ones that have are discussed here, extrapolating by inference to similar deposits in other basins. Comparisons can be made where detailed studies of the same facies have been made in one or more basins. This part of the book, then, is divided by "depositional environment." The associated variations and controls on the deposits of each environment, their diagnostic features, and the field-study techniques used to derive interpretable data from the outcrops constitute another focus of these chapters. Since the modern environments grade into one another and since their sediments interfinger in the ancient, overlaps occur in the subject material of the different chapters.

Many of the detailed papers on sedimentology concern the basins' lacustrine deposits, even though these facies volumetrically constitute only a small percentage of most of the basin fills. This disproportionate coverage may be due to their being one of the few fossiliferous deposits, to their containing relatively well-understood and well-preserved primary sedimentary structures, and to their being (often) widespread and stratigraphically correlatable within the basins; more recently, they have been the focus of attention as potential source rocks for hydrocarbon accumulations in the basins. In contrast, relatively few detailed sedimentological studies have been made of the mudstone and fluvial sandstone facies. The mudstone outcrops are often poorly exposed, vegetated, or eroded away, and they contain sedimentary structures that occur on a modest, easily overlooked scale.

The fluvial deposits often are just plain confusing. The record of fluvial sedimentation is not always cumulative. The deposits are commonly subject to postdepositional and syndepositional erosion, especially in the braided environments common in these basins. Although the patterns of sedimentary structures are not random, variable preservation of the deposits of each sedimentary event has often left an assemblage of deposits with ill-defined and partial trends. In addition, the study of modern fluvial environments is still evolving rapidly, and the complete range of variability of these deposits has not yet been defined. For these reasons, and because individual fluvial sandstones cannot be correlated stratigraphically, it is not surprising that many of the earlier studies concentrated on the more easily quantifiable petrologic characteristics of the fluvial sandstones.

In reviewing the relevant literature, one is presented with many opportunities for misquoting a reference or misusing the intent of the conclusions. The rarely quantified science of field geology has a long way to go before everyone writes exactly what he or she means to convey, and before everyone interprets written words in the same manner. Often enough, one runs across a reference in a given context and, upon looking up the reference, discovers that it referred only obliquely to the topic under discussion. Data from the literature must be used as carefully as data gathered directly from the rocks: they are subject to the same processes of filtering and interpretation. Use of the extant literature allows an individual a much broader perspective, but it also removes the geologist one step farther from the primary evidence.

Regional Setting

The processes of rifting that broke up Gondwana, creating the Triassic–Jurassic rift basins and finally the separate continental masses of North America, Greenland, Europe, and Africa, persisted from the late Paleozoic to the Early Tertiary (Van Houten 1977; Ziegler 1981, 1985). Rift basins are associated with nearly all phases of this system, but there was a major episode of rift-basin formation and filling during the Triassic and Early Jurassic.

In Greenland, in England, and possibly locally in Morocco, many of the same basins that were active in the Triassic have a history that extends back to the Permian, although not in the same extensional tectonic regime. In North America, the sedimentary record within most of the basins did not begin until the Late Triassic. Rifting in Greenland and England continued irregularly into the Early Tertiary before final continental separation took place in the northern North Atlantic. In contrast, rift-basin formation in North America and Morocco had ceased and the separation of North America from Africa had taken place by Middle Jurassic time. Only the Triassic–Early Jurassic rift-basin deposits are addressed in this book (figure II-1).

Whereas the onshore North American basins were located far inland, those of Europe and North Africa were located near the borders of Tethys and were subject to marine incursions. Eastern Greenland was situated close to the Arctic Ocean and was also subject to periodic marine influence. The sea-level curves of Vail, Mitchum, and Thompson (1977) and of Hallam (1984) suggest that eustatic sea level was generally rising during the Early and Middle Triassic, with a more rapid sea-level fall either in the Late Triassic (Hallam) or shortly after the beginning of the Jurassic (Vail, Mitchum, and Thompson 1977). As shown in chapter 10, however, the effects of changes in eustatic sea level on the sedimentation patterns of the rift basins—even those that were located near the seaways—are ambiguous.

Most of the rift basins are located along and parallel to the structural trend of the previous Variscan (or Hercynian) orogen. The axes of the basins commonly follow this trend rather than the prevailing direction of Triassic maximum principal horizontal stress, as would be expected on a homogene-

Figure II-1. Location of onshore rift basins on the margins of the Atlantic Ocean, with continents reconstructed to approximate their Late Triassic–Early Jurassic configuration. The numerous offshore basins are not shown. Basins discussed in the text are named. (After Van Houten 1977; Clemmensen 1980; and Glennie 1984)

ous basement. The associated basic dikes (figure II-2), however, apparently formed an array that was more sensitive to deep-seated contemporaneous stresses (King 1961; May 1971). Intrusive volcanic activity was widespread in North America, and extrusive basalts are found in the Moroccan and central North American provinces.

Early descriptions of the rifting process postulated broad thermal doming along the trend of the future continental separation. More recently, it has been suggested that regional subsidence accompanied the initial rifting processes, but that, once separation took place, the newly freed margins of the continents were probably uplifted (Watts 1981; Fairbridge 1979). Since the rift basins were formed prior to and synchronously with final continental separation (Schlee and Jansa 1981), many of the tectonically produced regional paleoslopes may have reversed during rift-basin sedimentation.

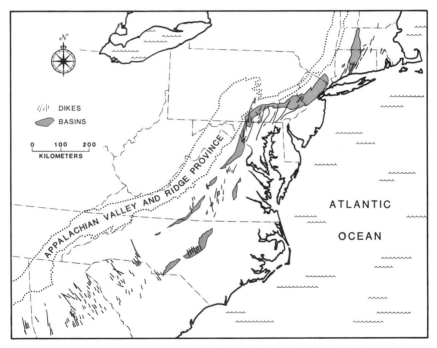

Figure II-2. Rift basins of North America inherited their orientation from the structures of the Appalachian orogen on which they lie. This congruence led to early ideas that the two were temporally and causally related. The orientations of the basaltic dikes, however, were apparently more sensitive to the deep-seated stresses associated with continental rifting—and possibly to the presence of a local hot spot (Philpotts and Martello 1986; May 1971). (After King 1961, Figure 4.1. Reproduced with the permission of the U.S. Geological Survey)

The origin of the rift basins was primarily extensional, although some large-scale wrench faulting may have occurred after Triassic time (Swanson 1982). Some authors have attributed the origin of the rift basins to continental-scale wrench faulting (Manspeizer, Puffer, and Couzminer, 1978; Manspeizer 1981), but Ratcliffe and Burton (1985) and Ratcliffe et al. (1986) have suggested that the evidence for wrench faulting, at least in the Newark basin, is misleading and can be explained by local structural complications in a more conventional, extensional tectonic regime.

A final note of regional importance to the rift-basin deposits relates to the Triassic–Early Jurassic climate. Robinson (1973) has reconstructed the global Late Triassic climate as one that was "dry year-round" in the interior areas of the supercontinent that were occupied by rift basins. Deposits in some of the rift basins suggest local variations, but the sedimentary record in the rift basins shows these sediments to be largely the product of an arid to semiarid climate.

5 / Alluvial-Fan Deposits

Conglomeratic deposits of alluvial-fan origin are present in all of the Triassic–Jurassic rift basins. In many basins, they form a significant portion of the basin fill, and they are usually the most distinctive and eye-catching deposits, locally containing clasts 2 meters or more in diameter. In other basins, the fan deposits are subordinate, forming only thin early-basin or narrow basin-margin sequences.

The existence of fan deposits is primarily a function of the local tectonic regime. In Triassic–Jurassic rifts, tectonism usually created basin-margin faults and large vertical offsets, resulting in significant topographic relief. Fans also formed along generally unfaulted basin margins, although these fans are still inferred to have originated in high-relief source areas. The high potential energy created by this relief brought sediments rapidly from the source terrane to the depositional basin, usually leaving them texturally and compositionally immature.

Climatic controls affected the deposits in more subtle ways, determining the relative percentages of different types of fan deposits. Fan morphology was determined by precipitation and river discharge rates, as well as by the rates of erosion and the consequent ratios of clay, sand, and cobbles produced in the source area. Tectonics, however—usually in the form of normal faulting—controlled the presence or absence of alluvial fans, their location, and their internal sequences.

The irregular pattern of tectonic/fault activity is often found recorded in the sedimentary patterns within the alluvial-fan conglomerates. Fining-upward patterns of tens or a few hundreds of meters indicate initial tectonic activity followed by quiescence and erosional lowering of the source area. Such patterns are often repeated several times within a basin. Fan deposits at

115

one end of a basin may be correlative with finer sediments at another end of the basin, even though they occur along the same border fault; this indicates irregular activity along the basin-margin fault.

Fans themselves are important controls on internal rift-basin sedimentation patterns. During high rates of deposition, fan sediments may overwhelm most other environments, in which case the basin-fill sequence is fan-dominated. Large fan deposits may also extend across the longitudinal drainage patterns of these narrow rift basins, and in some instances are inferred to have been the cause of local lacustrine conditions.

Alluvial-Fan Subenvironments

Various types of modern alluvial fans are known, ranging from those dominated by perennial fluvial processes to those dominated by ephemeral debris flows. Deposits showing an equal diversity are interpreted as fan sedimentation in the ancient. Most sedimentologists who have studied ancient fan deposits in detail recognize a tripartite division of proximal-fan, midfan, and distal-fan subenvironments, but not everyone seems to be describing the same fan subunits or the same diagnostic features.

Hardie, Smoot, and Eugster (1978) have recently recognized another distinctive and useful depositional subenvironment that often occurs at the distal margins of arid fans: sand flats. They named the process that forms these sand flats *sheetfloods*. This term further confuses the existing nonstandardized terminology. Bull (1972) has used *sheet-flooding* to describe different processes that are operative in midfan regions. Until the terminology becomes standardized, authors must define their usage of terms; in this book, a distinction is maintained between midfan sheetfloods and fan-toe sheetfloods. The process of sheet-flooding has been known to geomorphologists for some time, but it was rarely used to interpret sedimentary deposits in the rift basins until Hardie, Smoot, and Eugster published their work.

Most authors (such as Bull 1972) have recognized four basic processes and resulting deposits associated with alluvial fans. These, in roughly proximal to distal order, are as follows:

1. Sieve deposits. These form when sediment-laden water flows onto a porous fan surface and the water abruptly infiltrates into the fan, leaving behind its sediment load as angular, clast-supported conglomerates. Sieve deposits are rare in the Triassic–Jurassic rift basins and in most other fan deposits, since they tend to occur near the often-unpreserved fan apex and require special source terranes (Bull 1972).

2. Debris flows (including the finer-grained mud-flow category). These are common in both modern and ancient deposits, forming as high-viscosity laminar flows of entrained sediments of all sizes. The deposits are consequently unsorted and matrix-supported. They are common in proximal and midfan environments, decreasing in abundance downfan.

3. Fluvial deposits. These deposits may be formed by either ephemeral or perennial braided streams; in some fans, they represent the dominant sedimentary process. Fluvial deposits are discussed in chapter 6, their complexity warranting a section of their own. Fluvial deposits become increasingly common downfan.

4. Midfan sheetflood (Bull 1972) or braid-channel (Hardie, Smoot, and Eugster 1978) deposits. These are primarily composed of horizontal-planar bedded sandstones filling ill-defined shallow channels on extensive sandy reaches of the midfan and distal-fan areas. During flood events, flow from channels upfan becomes diffuse in this area, spreading its sediment across the fan. The flow may become rechanneled downfan.

5. Fan-toe sheetflood deposits. Hardie, Smoot, and Eugster (1978) described this additional environment and its deposits, which are common on fans associated with saline ephemeral lakes. These form sand flats at the toes of alluvial fans, usually where they adjoin the low-gradient playa floor. They are primarily deposits from decelerating, shallow, unconfined flows. They differ from midfan sheetfloods in the aspects of change of gradient and decelerating flow, and as a result they contain finer-grained sand that is deposited in climbing ripples and planar bedforms, and often in graded beds.

These are the processes (and their deposits) that have been sought in the Triassic–Jurassic rift-basin conglomerates. The search for them has been highly successful, beginning with the first recognition in the late 1800s and early 1900s that the generally coarse-grained, poorly sorted and poorly stratified deposits were nonmarine alluvial-fan deposits instead of marine shoreline sediments. As knowledge of the sedimentological details grew, so did recognition of them in the ancient deposits—for despite their occurrence in high-relief settings, alluvial fans are primarily depositional regimes. Little postdepositional erosion takes place, due to high subsidence rates and high sediment supply rates; therefore sedimentary structures (and often the entire bed that recorded each depositional event) are usually well-preserved (Bull 1972).

Although they are ubiquitous, only a few of the alluvial-fan conglomer-

Table 5-1. **Variable processes and deposits that have been recognized in alluvial-fan subenvironments. Note the overlap and differences that highlight the wide variation among alluvial fans.**

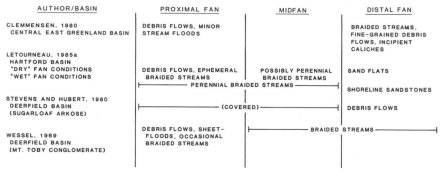

AUTHOR/BASIN	PROXIMAL FAN	MIDFAN	DISTAL FAN
CLEMMENSEN, 1980 CENTRAL EAST GREENLAND BASIN	DEBRIS FLOWS, MINOR STREAM FLOODS		BRAIDED STREAMS, FINE-GRAINED DEBRIS FLOWS, INCIPIENT CALICHES
LETOURNEAU, 1985a HARTFORD BASIN "DRY" FAN CONDITIONS "WET" FAN CONDITIONS	DEBRIS FLOWS, EPHEMERAL BRAIDED STREAMS	POSSIBLY PERENNIAL BRAIDED STREAMS	SAND FLATS
	├──── PERENNIAL BRAIDED STREAMS ────┤		SHORELINE SANDSTONES
STEVENS AND HUBERT, 1980 DEERFIELD BASIN (SUGARLOAF ARKOSE)	├────────── (COVERED) ──────────┤		DEBRIS FLOWS
WESSEL, 1969 DEERFIELD BASIN (MT. TOBY CONGLOMERATE)	DEBRIS FLOWS, SHEET-FLOODS, OCCASIONAL BRAIDED STREAMS	├──────── BRAIDED STREAMS ────────┤	

ates in the rift basins have been studied in enough detail to determine whether particular outcrops belong to proximal-fan, midfan, or distal-fan subenvironments, and most of these occur in the Hartford basin (table 5-1). Even within this one basin, significant variability exists. For instance, debris flows have been recognized within zones assigned to both the proximal fan and the distal fan. Hubert and Hyde (1982) have identified debris-flow deposits up to 5 kilometers from the inferred apex of a fan deposit. This suggests that the most important criterion to note in making divisions of fan deposits may be the relative percentage of each deposit at a given outcrop, or its relative grain size. In this respect, comparisons to similar deposits within the basin are more valuable than measurements against absolute standards distilled from numerous environments outside the basin.

Sieve and Debris-flow Deposits

Debris flows have been recognized in the rift basins on the basis of the following four general characteristics: matrix-supported clasts (figure 5-1); chaotic (or nonexistent) bedding, with unoriented clasts; coarse grain sizes, with exceptionally poor sorting; and planar to undulatory, unscoured lower surfaces.

In addition to these, several authors have used less common characteristics (such as those shown in table 5-2) to interpret deposits as debris flows. Debris flows are usually linear and move with pulses along their length; this explains their commonly hummocky upper surfaces, "wave-form clast fabrics," and lenticular cross sections. Two subtle yet informative characteristics that have been noted are the oriented micas in the fine-grained matrix, suggesting laminar flow, and the proportional relationship between clast size and bed thickness. Such proportional relationships have been noted by Arguden and Rodolfo (1986) in New Jersey, and by Bluck (1967) in Scotland. Large debris flows are capable of moving more material as well as larger clasts, and they were rarely subjected to reworking that would have decreased bed thickness while leaving larger clasts in place. Arguden and Rodolfo (1986, Figure 3) have plotted clast sizes against bed thicknesses, showing that both variables increase proportionately in the matrix-supported debris-flow conglomerates, whereas a maximum clast size is quickly reached in clast-supported conglomerates of a flood-stage fluvial facies; no clast-size to bed-thickness relationship is apparent in their braided-stream facies. Crude clast-size grading, where it exists in debris flows, is probably the result of slight turbulence within the flow and of the density and viscosity (a function of the percentage of water the flow contains) of the matrix of the flow.

More laterally extensive mud flows have been reconstructed in Scotland by Bluck (1967). These flows are also matrix-supported and sometimes graded, and they resemble other flows except in their unusually widespread nature. Interestingly, the degree of grading is good only in the medium-grained conglomerates; coarser deposits are poorly graded, and finer deposits show no grading—although their clasts sometimes display a "flow texture." The deposits are inferred to have been widespread viscous flows that were dispersed over the steep but planar surface of an alluvial fan. The wide lateral extent of the beds (scale not given) suggests that debris flows need not always be linear features.

Debris flows are created when flash flooding entrains erosion products that have accumulated near the site of erosion. They are therefore characteristic of fans in arid and semiarid climates. Where debris flows are not found in inferred fan deposits, their absence has been taken as evidence of more humid climates in which fluvial discharge was more regular and (as a result) where erosional debris did not accumulate between floods (Hubert, Gilchrist, and Reed 1982; LeTourneau 1985a, 1985b). There may be exceptions to this generalization, however, since modern debris flows are also known to form on some humid-climate fans (Nilsen, ed. 1985, p. 105).

Sieve deposits have not yet been recognized as such in rift-basin alluvial-fan deposits—or at least they have not been described in the published literature. Reynolds and Leavitt (1927) described large, unsorted, angular clasts in a local facies of the Mount Toby Formation in the Hartford basin as a scree or talus deposit, but they did not publish much sedimentological data. A suggestion that part of this formation might be composed of sieve deposits can be found in their description of the local absence of sand and silt matrix material interbedded with the clasts, although Wessel (1969) made no use of this model in his more recent study of these strata. Another possible example of a sieve deposit is shown in figure 5-2. The clast-supported fabric and angular clasts of this deposit contrast strikingly with the matrix-supported, rounded clasts of associated debris flows.

Midfan Fluvial and Sheetflood Deposits

The deposits of braided fluvial streams are often difficult to distinguish from midfan sheetflood accumulations in the sedimentary record, and sometimes are difficult to distinguish in the modern as well. A gradation of environments, from ephemeral to perennial braided streams, can be identified; and when the systems flood, partially reworked stream deposits become mixed with flood deposits, which may in turn be reworked by later fluvial processes. In addition, many braided stream processes and products are similar to those of midfan sheetfloods.

Figure 5-1. Debris-flow deposits: unsorted, matrix-supported conglomerates. (A) Lower Carboniferous clasts in a Triassic fanglomerate in South Wales (scale 25 centimeters). (Photo courtesy of B. J. Bluck) (B) Two finer-grained but still massive, structureless debris-flow deposits in the Kerrouchen basin, Morocco. Note that the lower flow is almost a mud flow, and that a caliche soil profile had been developing on it at the time of deposition of the upper flow. (Photo by the author)

Table 5-2. **Different criteria that have been used to identify rift-basin debris-flow or mud-flow deposits in outcrops. Note that some criteria are not compatible with others—a result of the high variability within different fan deposits.**

AUTHOR/BASIN	INTERNAL FABRIC							BED SHAPE				CLAST ORIENTATION				
	Matrix-supported clasts	No internal bedding or clast orientation	Poor sorting	Normal grading	Inverse grading	Large clasts at top and edges	Clast size proportional to bed thickness	Lenticular beds/irregular shape	Hummocky/irregular upper surface	Planar/undulatory nonscoured base	Persistent lenses	Clasts parallel to flow	Clasts normal to flow	Clasts with "wave-form" orientation	Mica flakes parallel to flow	Imbricated clasts
BLUCK, 1965 SOUTH WALES	×	×			×							×	×	×		
CLEMMENSEN, 1980 EAST GREENLAND		×							×							
JONES, 1975 ESSAOUIRA		×	×					×	×							
LETOURNEAU, 1985a,c HARTFORD	×	×	×		×	×		×	×							
LINDHOLM et al, 1979 CULPEPER	×	×							×					×		
STEEL, 1974 SCOTLAND	×	×	×	×		×			×			×				
STEVENS AND HUBERT, 1980 DEERFIELD	×	×		×					×			×	×			
WESSEL, 1969 DEERFIELD		×	×					×				×	×			

Table 5-3 lists many of the characteristics that have been used in detailed studies of Triassic–Jurassic alluvial fans to distinguish fluvial/sheetflood deposits from deposits of other environments. Again, the most promising approach is to compare individual deposits within a given basin. In general, these deposits are somewhat better sorted than associated nonfluvial accumulations, and they often display flowing-water features such as imbricate clasts (figure 5-3), crossbeds, silt drapes, and (rarely) antidunes (Hand, Wessel, and Hayes, 1969). Unlike debris flows, many of these beds have erosive bases. Stevens and Hubert (1980) noted that deposits they interpret as having formed in braided rivers are composed primarily (65 percent) of horizontal-parallel bedding, with another 25 percent being crossbedded. Shallow,

Figure 5-2. Clast-supported conglomerate: possible sieve deposit in rift-basin strata in South Wales. Note the angular clasts. The coin for scale is 2 centimeters in diameter. (Photo courtesy of B. J. Bluck)

Table 5-3. Different criteria that have been used to identify rift-basin midfan sheetflood and braided-stream deposits in outcrops.

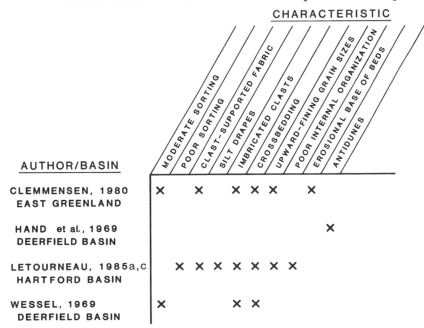

AUTHOR/BASIN	MODERATE SORTING	POOR SORTING	CLAST-SUPPORTED FABRIC	SILT DRAPES	IMBRICATED CLASTS	CROSSBEDDING	UPWARD-FINING GRAIN SIZES	POOR INTERNAL ORGANIZATION	EROSIONAL BASE OF BEDS	ANTIDUNES
CLEMMENSEN, 1980 EAST GREENLAND	X		X	X	X		X			
HAND et al., 1969 DEERFIELD BASIN								X		
LETOURNEAU, 1985a,c HARTFORD BASIN		X	X	X	X	X	X	X		
WESSEL, 1969 DEERFIELD BASIN	X				X	X				

Figure 5-3. Imbricated clasts and internal structure within a conglomerate, probably of midfan braided-stream/sheetflood origin, Essaouira basin, Morocco (hammer for scale). (Photo courtesy of F. B. Van Houten)

unchanneled sheetfloods across fan surfaces can also deposit thin, graded beds of very coarse material (figure 5-4).

In another study, Hubert, Gilchrist, and Reed (1982) suggested that deposits in the Portland Arkose that were 79 percent horizontal-parallel bedded and had formed laterally continuous sheetlike sand bodies (figure 5-5) were the record of wide, shallow "sand-bed rivers subject to repeated floods." Each 1- to 6-meter-thick bed with a basal shallow scour has been interpreted as the record of a single depositional event, with pulsating flow being implied by occasional internal mud drapes. Hubert, Gilchrist, and Reed did not interpret the position of these deposits on a fan or floodplain, but the rocks occur in outcrops 3 to 6 kilometers from the border fault and may be midfan sheetflood deposits. Hubert and his co-workers used the thick horizontal-planar bedded sands of the modern Bijou Creek catastrophic flood deposits (McKee, Crosby, and Berryhill 1967) as a possible modern analogue. Although these modern sediments occur significantly farther from high-relief source areas, their internal sedimentary structures are similar to those described for this part of the Portland Arkose.

Conglomerates in Scotland inferred by Bluck to have originated in an

Figure 5-4. Series of thin, graded, sheetflow deposits within coarse-grained, proximal, alluvial-fan deposits, Hartford basin, Connecticut. (Photo by the author)

"alluvial-fan stream" environment may be either structureless or crossbedded; they are predominantly clast-supported (Bluck 1967). These conglomerates have a smaller range of clast-size distribution than associated debris-flow conglomerates do, reflecting either a change in the competence of the transporting medium or a greater transport distance. In this basin, however, clast size in the stream deposits is proportional to bed thickness, possibly indicating little or no reworking of older, coarser underlying conglomerates. Despite this, the stream conglomerates are often found in broad channels (up to 12 meters wide), but Bluck suggested that—since the crossbedded conglomerates commonly extend beyond the limits of the channels—the erosional events that created the channels were unrelated to the subsequent depositional processes.

Fan-toe Sheetflood Deposits

The sedimentology of fan-toe sheetfloods has only recently been studied in modern alluvial-fan environments, and few papers have used this model as

Figure 5-5. Coarse sandstone and conglomerate of probable midfan sheetflood origin, Hartford basin, Connecticut. Note the horizontal-planar and low-angle-planar bedding, and the crossbedding. Note also the shallow pocket of conglomerate to the left of the hammer—probably a shallow channel within the overall sandy-bedded, widespread, multiple-channel sheetflood. (Photo by the author)

yet. Although the fan-toe sheetflood process could arguably be assigned to the playa-lake environment, it is included in this chapter because the fan-toe sand flat is the final sandy deposit of alluvial-fan floodwaters.

Hubert and Hyde (1982) used the sand-flat/sheetflood model to reinterpret parts of the Blomidon Formation in the Fundy basin. This formation had previously been interpreted as a primarily lacustrine deposit because of the presence of fish fossils, "even, uniformly thick, laterally persistent stratification, rhythmic lamination, thin lamination, graded bedding, disturbed stratification, and oscillation ripple marks" (Klein 1962*b*, p. 1136). One lithofacies was inferred to be a below-wave-base deposit, while a lithofacies that showed evidence of currents and subaerial exposure was a "periodically uncovered shallow lacustrine shelf" (p. 1136). Hubert and Hyde reinterpreted both lithofacies as being part of a playa complex with alternate subaqueous and subaerial phases, and they reinterpreted those deposits within it that displayed graded bedding (figure 5-6) as sheetflood ("sheetflow") deposits on the fan-toe sand flats.

Figure 5-6. Graded bedding in fan-toe sheetflood deposits, Fundy basin, Canada. (A) Cyclic sequences of graded beds (G) fining upward into sandy mudstone beds (M). (B) Three graded beds from the lower part of a cycle of the type depicted in (A) (coin for scale). (Photos courtesy of J. F. Hubert, from Hubert and Hyde 1982, Figure 3. Reprinted with the permission of *Sedimentology*)

Hubert and Hyde defined nine types of sequences within the graded beds, all of them with fining-upward grain sizes and with sedimentary structures that indicate decreasing velocity of the depositing floodwaters. The grain sizes range from sand to mud, and the sedimentary structures include crossbeds, ripple and climbing-ripple cross-stratification, and horizontal-planar bedding that usually occurs in capping mudstones. The beds are less than 1 meter thick and have a mean thickness of only about 10 centimeters.

Hubert and Hyde (1982, p. 463) envisioned the depositional process as one in which "the flow velocity and flow depth [of a flash flood] abruptly drop where the 3°–6° slope of the fan rapidly changes to the less than 1° slope of the sandflat. Deceleration of the sheet flow over the sandflat leads to deposition of . . . graded beds." Deceleration and decrease in flow depth were also caused by the dispersal of channeled flow from the fan onto the unchanneled sand flats.

In sum, thin, extensive, graded beds that contain a variety of combinations of sedimentary structures indicative of decelerating flow have been interpreted as sand flats deposited by sheetfloods at the junction of alluvial fans and playa-lake flats. In the Hartford basin, thin, horizontal-planar bedded sandstone units of probable fan-toe sheetflood origin (figure 5-7) are interbedded with the black mudstone deposits of more permanent lakes.

In England, Henson (1970), Tucker (1977), and Tucker and Burchette (1977) used the earlier, geomorphic, desert process models of Cook and Warren (1973) and Blissenbach (1954) to interpret similar deposits. Thin (3 to 15 centimeters thick) yet extensive (up to 50 meters) graded pebbly sandstone beds (figure 5-8) were inferred to be sheetflood deposits. The units have sharp bases with minor scour relief and local pebble lags. They contain internal crossbeds, and commonly have current ripples at the top. Vertebrate tracks and desiccation cracks are often present on the upper surfaces. The sheetflood beds are interbedded with the deposits of "temporary muddy lakes"—mudstone to fine sandstone units with horizontal laminations, oscillation ripples, mud cracks, and raindrop impressions.

These sheetflood deposits represent the transitional facies between alluvial fans and the playa floor. Although the British deposits were described (and the sheetflood model was used) prior to Hardie, Smoot, and Eugster's (1978) publication of a sedimentological description of modern analogues, the British authors' use of models of geomorphic processes was not widely followed by American geologists. This may have been because the main thrust of Tucker and Burchette's study was the associated vertebrate tracks, while the sheetflood model was only a part of Tucker's paper, and thus was easily overlooked.

Figure 5-7. Thin horizontal-planar bedded sandstones: fan-toe sheetflood deposits interbedded with lacustrine deposits, Hartford basin, Connecticut. (A) Vertical cut showing thin bedding (hammer for scale). (B) Bedding plane showing parting lineation (pencil for scale). (Both photos by the author)

Figure 5-8. Thin, laterally extensive, graded, pebbly sandstone beds (sheetflood deposits) interbedded with laminated, wave-rippled, and desiccated siltstones (playa deposits), South Wales (hammer for scale). (Photo courtesy of M. E. Tucker. From Tucker 1977, Figure 13c. Reprinted with the permission of the *Geological Journal*, copyright 1977, John Wiley & Sons)

Nonstandard Alluvial Fans

The alluvial-fan deposits described so far have been those of conventional, arid, high-gradient, fault-bounded fans. The rift-basin setting has also produced several other types of fan deposits. Alluvial-fan deposits have been documented along the unfaulted margins of some rift basins and at the ends of others where they prograded along the axis of the basin. Moreover, syndepositional or postdepositional border-fault migration can produce complications in the vertical facies sequences.

In the High Atlas region of Morocco, several rift basins are aligned along a major pre-Triassic structural trend (Van Houten 1977). The Tichka Massif or Massif Ancien, an upper Paleozoic intrusive complex that created a major Triassic topographic high within the structural trend, divided the basins into eastern and western provinces and provided the primary source of coarse basin-fill sediments. Thus the rift-basin alluvial fans prograded eastward and westward into the basins primarily from the ends, rather than from the faulted margins of the basins (figure 5-9).

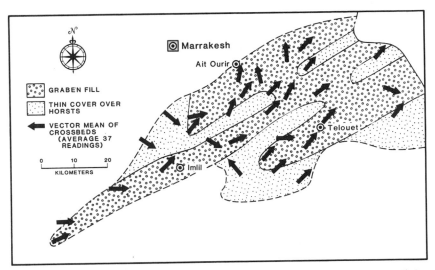

Figure 5-9. Paleocurrent dispersal patterns in the Central-N'Fis basin area (High Atlas, Morocco), showing axial drainage of the basin derived from the main alluvial-fan facies at the southwestern end of the basin. (After Mattis 1977, Figure 1. Reproduced with the permission of the *Journal of Sedimentary Petrology*)

Brown (1980) noted that the the basal conglomeratic lithologies (of two distinct fan complexes) in the western (Essaouira) basin of Morocco are confined to the central graben, in a structural setting where the basin consists of a series of parallel horst and graben blocks, all of which subsided independently but concurrently. The conglomerates are not fault-bounded at their proximal ends; instead, they "overlie an erosional front of deposits that once onlapped" the source area at the end of the basin (Brown 1980, p. 991). Paleodrainage was almost entirely longitudinal, westward down the axis of the basin.

Mattis (1977) and Jones (1975) have documented easterly paleodrainage down the axis of the structurally similar Central-N'Fis rift basin complex on the other side of the Massif Ancien (figure 5-9). The main fan deposits thin away from this end-of-the-basin source area, and only minor fan deposits are present along the faulted margins of the internal horsts. Mattis (1977, p. 118) wrote that "the lack of major conglomerates along the margins . . . indicates that there was little topographic relief at borders of the Central High Atlas Triassic basins."

In another area, alluvial-fan deposits on opposite edges of the Newark basin in New Jersey provide an interesting contrast to one another. The Hammer Creek Conglomerate on the northwestern side of the basin is a "standard" faulted-margin alluvial-fan deposit. Much of the sandy to conglomeratic Stockton Formation, located along the generally unfaulted southwestern margin of the Newark Basin (opposite the Hammer Creek Conglomerate), may also be classified as an alluvial-fan deposit (Glaeser 1966; Turner-Peterson 1980). The environment of the latter has been interpreted as one in which fans were dominated by fluvial processes—an environment of "well-established streams" on laterally coalesced fans with relatively gentle slopes (Turner-Peterson 1980).

Because it lacks an associated basin-margin fault, the Stockton Formation offers an example of a fan deposit that thickens into the basin, rather than being thickest near a rapidly subsiding basin margin. It differs markedly from the high-gradient, Hammer Creek debris-flow deposits. Both formations were deposited under the same climatic conditions, but the difference in tectonic regime produced significantly different sedimentary processes and quite different deposits.

Randazzo and Copeland (1976) and Randazzo, Swe, and Wheeler (1970) have described an interesting situation in the Wadesboro basin of North Carolina. In this basin, there are no conglomerates within the thickest basin-fill sediments that occur against the major basin-margin fault, yet there are conglomerates in the thinner deposits on the opposite side of the basin. Randazzo, Swe, and Wheeler have suggested that postdepositional faulting has bisected the basin longitudinally, uplifting and destroying the half of the basin that contained the original border-fault conglomerates, and juxtaposing finer-grained distal facies against a postdepositional fault.

Wessel (1969) and Willard (1951, 1952) have studied the Mount Toby Formation of the Deerfield basin and have shown that it is relatively thin toward the present (faulted) eastern boundary of the basin. This is apparently because the Mount Toby Formation consists of sediments shed onto a fan from a younger fault system that was stepped back away from the basin and the initial basin-margin fault. The initial fault became inactive and was subsequently buried by relatively thin fan deposits shed from the secondary, younger fault system.

Facies: Patterns and Controls

Alluvial-fan deposits may interfinger distally with deposits of a variety of environments; thus they may be overlain or underlain by those same deposits. In rift basins, they grade distally into fluvial, paludal, playa, lacustrine, and even marine-evaporite deposits. The nature of the sedimentary interaction between fans and other lithologies can provide information on the conditions of deposition.

Gilchrist (1979) and LeTourneau (1985a, 1985c) have suggested that, in the Hartford basin, fan deposits without debris flows—fans that graded out into shoreline sandstones and lacustrine shales—are the products of relatively humid climates. Within the same formation, associated fan deposits that contain debris flows and that grade distally into sand-flat and playa or fluvial floodplain sediments reflect more arid conditions. Fans that developed during relatively arid phases are inferred to have had steeper gradients but more limited areal extents than did those that originated during less arid climatic phases. Moreover, fan morphology and size may have been controlled in part by the intensity of the basin-margin faulting. By analogy to modern fans in Death Valley, California, LeTourneau and McDonald (1985) have proposed that small, radial-shaped fans were produced by rapid subsidence and active faulting, whereas inactive faulting allowed the formation of larger, coalescing fan complexes. The widespread, relatively fine-grained fan deposits of the Stockton Arkose in the Newark basin were formed at an essentially unfaulted basin margin and may owe much of their broad, low-slope morphology to a concomitant lack of associated tectonism.

Bluck (1965) has inferred that climate also controlled the variability of Triassic fan deposits in South Wales, where he identified two types of fan deposits in one formation. The fans interpreted as products of a drier climate contain debris flows with poor sorting, poorly defined bedding, and structureless sediments. The plan-view contours of the maximum clast size in these deposits have an irregular lobate pattern (figure 5-10B)—the lobes possibly being associated with the main paths of debris-flow surges of coarse material onto the fan. The wetter-climate fans have more regular maximum-clast-size contours (figure 5-10A), with only a few minor irregularities in contours that correlate with the major distributary channel systems. These deposits are better sorted and have more definitive sedimentary structures.

Small-scale (meters to a few tens of meters) vertical patterns in alluvial-fan sediments have been interpreted as records of autocyclic processes on fans. Hubert and Hyde (1982) have observed fining-upward sedimentary sequences and have inferred them to be the product of channel avulsion on the fans. In some cases, the sequences include an initial thin coarsening-upward phase that records the early progradation of sediments into a new area.

Most of the vertical patterns seen in alluvial-fan deposits, however, reflect tectonic activity. Fining-upward patterns of tens to a few hundreds of meters thickness (figure 5-11) are commonly interpreted as evidence of erosional reduction of topographic relief in the source area during temporary tectonic inactivity.

These fining-upward trends may also be observed within a series of related deposits, such as the 60- to 70-meter sequences of sand-flat deposits

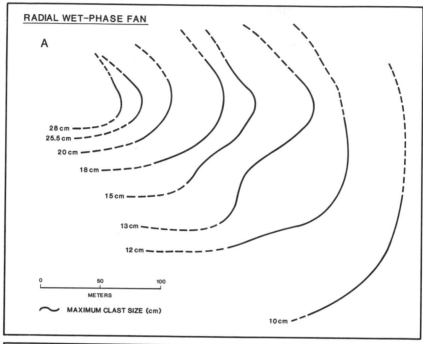

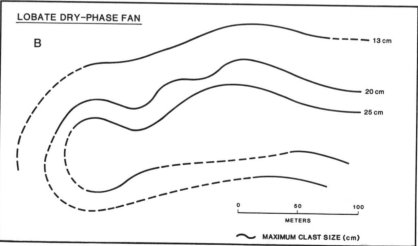

Figure 5-10. Contours of maximum clast size differ in shape for small rift-basin alluvial fans in South Wales. The radial pattern (A) is associated with crossbedded sandstones and is interpreted as being the product of a more humid climate. The lobate pattern (B) is associated with debris-flow deposits and is interpreted as being the product of a more arid climate. (After Bluck 1965, Figures 4 and 7. Reproduced with the permission of *Sedimentology*)

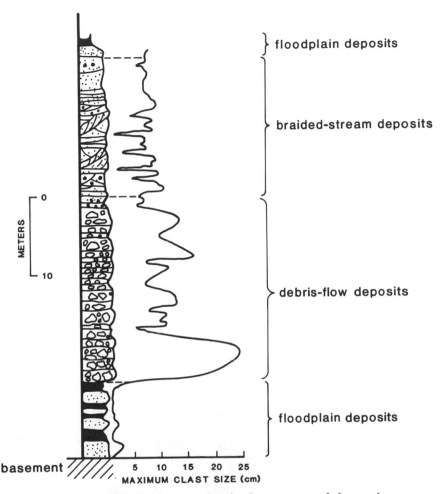

Figure 5-11. Large-scale fining-upward grain-size patterns and changes in depositional environments (Triassic rift-basin sedimentary rocks in Scotland), interpreted as representing "geomorphic maturing of the basin margins" during an era of tectonic stability. (After Steel 1974, Figure 14. Reproduced with the permission of the *Journal of Sedimentary Petrology*)

that are more thinly bedded toward the top, as noted by Hubert and Hyde (1982). Finally, such trends may be expressed as vertical progressions from proximal-fan through more distal-fan environments—that is, in the debris-flow to stream-flood to braided-stream sequences noted by Bluck (1967) and Steel (1974) in Scotland. In the former study, the relatively distal facies were documented to have onlapped the proximal facies and even parts of the source area.

Another possible indication of syndepositional tectonics is the local presence of low-angle unconformities within alluvial fans and associated deposits. Hubert and Hyde (1982, p. 469) noted a 7-degree angular unconformity at one outcrop, attributing it to "abrupt sinking of the valley floor." This interpretation would be especially attractive if a similar unconformity can be found at the same horizon at several adjacent outcrops, below a reintroduction of coarser material in the basin. A possible example of this can be seen in figure 5-12.

Such intrabasinal unconformities may be extensive; at least they need not be restricted to the fan deposits in the vicinity of the major basin-margin faults. Schutz (1956) has suggested that an unconformity exists within the early basin-fill strata in the Pomperaug outlier of the Hartford basin, 40 to 50 kilometers west of the main border fault. It was inferred on the basis of an abrupt 20-degree change in dip and 10-degree change in strike within the limited exposures along the Pomperaug River. On the other hand, Arguden and Rodolfo (1986) have suggested that tectonic uplift of isolated portions of the Newark basin margin caused local fan reactivation that did not have a counterpart elsewhere in the basin.

Figure 5-12. Possible intraformational angular unconformity (plane between the arrows) caused by renewed tectonism, Hartford basin, Connecticut. Note the angular relationship and the sudden increase in grain size of the sediments above the unconformity. (Photo by the author)

Summary

Alluvial-fan sediments are probably the most characteristic deposits of rift basin settings. While a wide variety of conglomeratic deposits exists—some of which are still unexplained—an examination of these rocks is a good way to begin the second part of this book. Alluvial-fan sediments have a high preservation potential, so most of the processes and deposits that have been documented in modern environments can be recognized in the ancient record.

In some respects, alluvial-fan sediments are also easy to study, having unambiguous diagnostic characteristics such as debris flows that are unique to fan-depositional environments. Although much remains to be learned about the occurrence and distribution of different subenvironments on alluvial fans, the depositional models derived from studies of modern alluvial-fan environments lend themselves nicely to the interpretation of Triassic–Jurassic rift-basin strata. At present, the major discrepancies between the models and the strata are the rarity of recognizable sieve deposits in the ancient record and the ambiguities between midfan sheetflood and fluvial environments. Other differences are noted in chapter 12.

Data Collection and Interpretation Techniques Illustrated by Studies of Rift-Basin Alluvial-Fan Deposits

A list of the techniques that have been successfully used in the field study and interpretation of alluvial-fan deposits in the Triassic–Jurassic rift basins must start with recognition of the distinctive sedimentary characteristics of the different subenvironments that are present in fan deposits. These are usually straightforward and are listed in many texts. It also helps to be open to other opportunistic identifying criteria that are not always listed; an example of such a criterion might be oriented mica flakes (Lindholm, Hazlett, and Fagin 1979), suggestive of the laminar flow within advancing debris flows.

Interpretations of the presence or absence of these deposits, as well as of their lateral and vertical distribution, help to define paleogeography and provide insights on paleoclimate. Associated lithologies and their inferred depositional regimes aid in the overall understanding of the fan deposits, but they must be interpreted carefully: fan environments respond first and foremost to tectonic/structural controls, whereas other environments respond more equally to a combination of tectonics, climate, and other phenomena such as eustatic sea level.

Plots of clast-size trends, in conjunction with regional patterns of crossbedding azimuths from associated sandstones (figure 5-13), have been useful in a

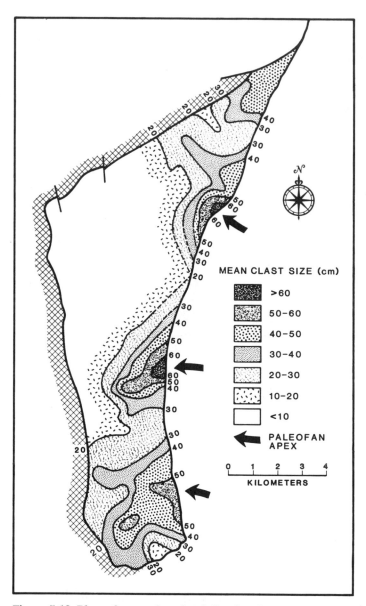

Figure 5-13. Plots of mean clast size define fan shape, size, and position in the Deerfield basin, Massachusetts. Note the larger scale in comparison to Bluck's small fans, and note that mean rather than largest clast size is plotted here. (After Wessel 1969, Figure 30. Reproduced with the permission of the Dept. of Geology and Geography, University of Massachusetts)

number of studies in helping to determine the location and extent of specific ancient alluvial fans. Bluck (1965) showed that the rate of change of clast sizes downfan in South Devon is an exponential function, as it is in modern fans, and that it correlates well with the rate of decrease in bedding thickness. He felt that there was too much variability in modern fan slopes, however, so he did not attempt to apply his rate of decrease in clast and bed sizes in the ancient deposits to the reconstruction of fan slopes. Measurements of clast imbrication have also been useful, but plots of the orientations of clasts' long axes have been less so, often yielding ambiguous patterns.

A unique argument was used by Bluck (1967) in reconstructing the local basin tectonics in Scotland. As in South Devon, the maximum clast size of a debris-flow conglomerate proved to be proportional to the thickness of the bed across the area studied; this indicated that little if any erosion and reworking of these deposits had occurred. Bluck reasoned that, since this was so, the bedding thickness should increase as clasts became larger toward the source area. If bedding thicknesses increased in that direction, the entire formation should become thicker toward the source, and therefore the unexposed basin margin must be a faulted one in order for it to have accommodated a formation with this geometry.

Patterns of the local paleotectonic history of different basins are commonly inferred from vertical trends of facies patterns and grain sizes. Studies of this sort have probably not been taken nearly as far as they could be. Sedimentologists who have access to this type of information are not always interested in making regional tectonic inferences or correlations from it. Care must, of course, be taken to separate climatic patterns from tectonic patterns. For example, Arguden and Rodolfo (1986) suggest that the pattern of relatively fine-grained, alluvial-fan facies assemblages in the Newark basin was predominantly controlled by climate during early stages of deposition. However, renewed tectonism subsequently overwhelmed the climatic controls, creating thick, tectonically produced conglomerates that responded to climatic control only secondarily.

6 / Fluvial Deposits

Like alluvial-fan deposits, fluvial deposits are common in almost all of the rift basins. Because of the coarse sediment and generally high gradients of slopes within the basins during fluvial deposition, most of the rivers were variants of braided systems, although several high-sinuosity fluvial regimes have also been described.

Coarse, sandy rocks in the basins were recognized in a gross sense as having been fluvially derived at about the same time that the alluvial-fan deposits were first properly identified. The subdiscipline of fluvial sedimentology was only beginning to be developed in the late 1940s and 1950s, however, and the full significance of the internal patterns of sedimentary structures within fluvial deposits was not appreciated until the late 1950s and early 1960s (Miall 1978a). Moreover, it was not until 1962 that deposits of a modern braided fluvial system were first described, at last giving geologists working on ancient fluvial deposits relevant models to apply to their rocks.

The nature of fluvial deposits is controlled by a combination of tectonics, climate, and source terrane. These affect the size and volume of sediment available to the fluvial system and the volume and variability of the fluvial discharge. In most Triassic–Jurassic basins, tectonics created high-relief source terranes that provided abundant coarse sediment to the basin interior. In the generally arid climates, fluvial discharge was irregular. Frequent flooding and steep gradients provided currents that were competent to transport very coarse sediment.

As was noted in chapter 5, fluvial deposits are one of the principal components of alluvial-fan sequences. Fluvial action also produces lower-gradient, distal-floodplain sequences that are composed of both channel and overbank deposits. The following discussion is divided into sections on braided, meandering, and overbank fluvial deposits.

Braided-river Deposits

Because of variability in discharge, the deposits of braided streams are often subject to reworking, and a single sandstone bed usually records a number of depositional events. Thus, unlike the deposits of many meandering rivers, the vertical sequences of sedimentary structures in braided-river deposits often seem chaotic (figure 6-1). Although there is some order to these structures, a statistical Markov chain analysis is usually required to reveal it, and even then the resulting predictive sequence is a probabilistic one that can be applied only to the suite of deposits from which it was derived.

Generalized braided-river models have been created on the basis of modern deposits, but with each modern stream studied, it seems that another generic model is added to the list; as a result, models have proliferated to the point where they are no longer generic. This has left geologists who are working on ancient sediments the option of choosing one from a variety of specific models or using different but compatible components from the various models to reconstruct the depositional fluvial regime.

Table 6-1 lists characteristics of coarse-grained deposits that have been used by different authors as criteria for the recognition of braided-river deposits. This table is misleading in that it mixes the deposits of a number of obviously different braided fluvial regimes, but the variability of recognized sedimentary structures highlights the variability in braided fluvial systems. On the other hand, authors do not always mention all of the characteristics present in the outcrop they are studying, and there is probably more overlap in the suites of structures than is evident from the table. Rather than attaching special significance to one or two structures, most authors have used assemblages and the arrangement of sedimentary structures as diagnostic evidence of braided fluvial environments.

Using outcrop evidence, several authors have suggested that bars were present in the Triassic–Jurassic rivers. Reconstruction of bar types allows for the interpretation of paleoriver geometry and ultimately of paleogeography. To date, bar reconstructions have centered on the association of horizontal-planar bedding (interpreted as bar backslopes) with tabular crossbeds (interpreted to be downstream slipfaces of the bars). These interpretations can be found in the three different rift basins studied by Steel and Thompson (1983), Nadon and Middleton (1985), and Hubert et al. (1978). The last two

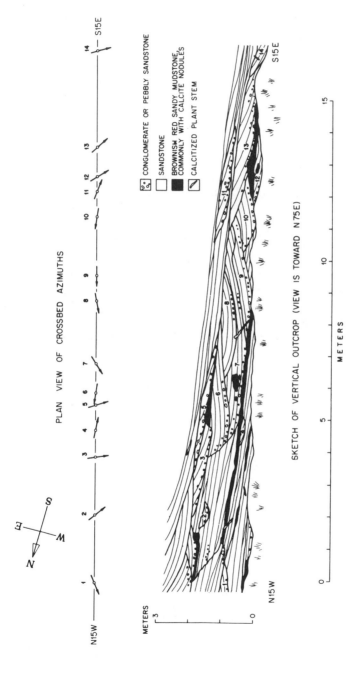

PLAN VIEW OF CROSSBED AZIMUTHS

METERS

SKETCH OF VERTICAL OUTCROP (VIEW IS TOWARD N 75E)

METERS

CONGLOMERATE OR PEBBLY SANDSTONE

SANDSTONE

BROWNISH RED SANDY MUDSTONE, COMMONLY WITH CALCITE NODULES

CALCITIZED PLANT STEM

Figure 6-1. Fluvial braid-bar complex in conglomeratic sandstone, Hartford basin, Connecticut. Note the complexity of the deposits, caused by multiple depositional events and numerous internal scours. Numbers in the sketch correspond to crossbed azimuths plotted at the top of the sketch. (From Hubert et al. 1978, Figure 8. Reprinted with the permission of the Dept. of Geology and Geography, University of Massachusetts)

Table 6-1. **Examples of some of the criteria that have been used to identify rift-basin braided-stream deposits in outcrops. Note the high degree of variability and mutually exclusive categories, reflecting the heterogeneity of this environment.**

CHARACTERISTICS

AUTHOR/LOCALITY	SEDIMENTARY STRUCTURES									ROCK FABRIC							EXTERNAL SHAPE	OTHER				
	ERRATIC INTERNAL ORGANIZATION	SILTSTONE DRAPES	RIP-UP CLASTS	HORIZONTAL-PLANAR BEDDING	TROUGH CROSSBEDDING	PLANAR-TABULAR CROSSBEDDING	IMBRICATED CLASTS	FINING-UPWARD GRAIN SIZES	COARSENING-UPWARD GRAIN SIZES	CLAST-SUPPORTED	MATRIX-SUPPORTED	WELL-SORTED	MODERATELY SORTED	POORLY SORTED	RAPID LATERAL TEXTURE VARIATION	BASAL AND INTERNAL SCOURS	SHEETLIKE SAND BODIES	BIMODAL CROSSBEDDING	LOW-VARIANCE CROSSBEDDING	CLASTS WITH PERCUSSION MARKS	FINER-GRAINED THAN FAN DEPOSITS	THINNING-UPWARD CROSSBED COSETS
CLEMMENSEN, 1980 EAST GREENLAND				×		×										×						
HUBERT et al., 1978 HARTFORD BASIN	×	×	×	×	×	×										×						
HUBERT AND FORLENZA, 1986 FUNDY BASIN				×	×	×			×							×	×			×		
LETOURNEAU, 1985a,c HARTFORD BASIN: EPHEMERAL	×	×					×		×													
PERENNIAL				×	×		×		×		×											
NADON AND MIDDLETON, 1985 FUNDY BASIN						×											×					
STEEL, 1974 SCOTLAND						×			×				×	×	×	×	×			×		
STEEL AND THOMPSON, 1983 ENGLAND	×					×	×		×	×		×	×			×	×					
STEVENS AND HUBERT, 1980 DEERFIELD BASIN	×	×		×	×				×							×	×					

papers also note interbedded trough crossbeds, which they interpreted to have been formed in interbar channels.

Actual bar morphology is more difficult to reconstruct. Hubert et al. (1978, p. 20) wrote that the "presence of gravel in the bedload of the river suggests longitudinal bars of diamond shape" in the lower part of the New Haven Arkose in the Hartford basin. The authors also suggested (p. 63–64) that the finer-grained sediments and associated scour surfaces higher in the section may indicate bars with a lingoid form.

Reconstruction of the types of bars that gave rise to existing sedimentary structures is difficult, given partial preservation and the vertical, two-dimensional exposures common to outcrops. In addition, a number of different types of bars are capable of constructing similar sedimentary structures. The spatial arrangement of the different structures, rather than their presence or absence, provides the best evidence for original bar and channel morphologies, but such data are commonly absent, and reconstruction of bar morphology has rarely been done convincingly in the rift-basin deposits.

The use of specific modern braided-river analogues has also been infrequent in the rift-basin interpretations. Nadon and Middleton (1985) suggested that the presence of conglomerates and a fining-upward trend in some of the fluvial deposits of the Fundy basin might indicate a Scott-type braided river, whereas associated sandier deposits may represent a Donjek-type (types as described by Miall 1978*b*); but again, point-for-point correlations of outcrop characteristics and modern river-specific models have rarely worked well.

Where outcrops permit observation of broad-scale characteristics, braided-river deposits can often be recognized in general by laterally extensive sandstones, indicating the typically high width-to-depth ratios of braided-river channels (figure 6-2). Such sand bodies commonly display internal scouring and reworking of earlier deposits, produced by multiple depositional events; they are usually associated with minimum volumes of fine-grained overbank material.

The syndepositional erosion within braided fluvial systems is reflected in crossbedded conglomerates in Scotland, where Bluck (1967) noted that clast sizes were no longer proportional to bed thickness (as they had been in the debris-flow deposits). Reworking of coarser, older deposits by braided streams resulted in large clasts that were incorporated into the braided-river deposits

Figure 6-2. Sheetlike sandstones with low volumes of fine-grained overbank material, typical of braided-stream deposits, Hartford basin, Connecticut. The complex depicted in figure 6-1 comes from this outcrop. (Photo by the author)

as apparent anomalies. The scouring also produced steep-sided remnant "pillars" (up to 2 meters high) of in situ overbank material within the conglomerates; and in conjunction with lateral migration of the fluvial systems, it created laterally extensive channel-base lag deposits. The meter-deep erosional hollows at the base of the conglomerates were often infilled by conglomerates that are notably finer-grained at the downstream end of the hollow.

Despite the heterogeneity of these deposits, vertical trends in braided-river deposits have been recognized in a number of examples. Fining-upward grain sizes have been noted in individual event beds by LeTourneau (1985a, 1985c), Steel (1974), and Stevens and Hubert (1980). They have been attributed to waning discharge and aggradation of stream beds in the later stages of flooding in ephemeral systems. Stevens and Hubert have also described extensive (outcrop-scale) scours at the bases of braided-river sandstones, which they attribute to lateral migration of braided rivers.

Fining-upward grain-size trends and sequences of sedimentary structures have also been extracted from some multistoried sandstone beds. Clemmensen (1980), Steel (1974), and Nadon and Middleton (1985) have briefly described such sequences from deposits that they have interpreted as originating from braided rivers.

More recently, Hubert et al. (1983), Hubert (1985), and Hubert and Forlenza (1987) have published some of the more detailed studies of braided fluvial deposits in the rift basins. Their material suggests that definitive vertical sequences exist in some of the Fundy basin's braided-river deposits. These "modal" patterns were derived through examination of measured sections and through Markov chain analysis.

The cyclic, fining-upward fluvial deposits described by Hubert and Forlenza (1987) occur in two lithofacies—one composed of sandstone and conglomerate, and the other composed primarily of sandstone. Statistically reconstructed modal cycles in the coarser deposits are 2 to 10 meters thick. They overlie a scour surface and consist of basal-trough crossbeds that decrease in thickness upward and then grade into horizontal-planar laminations with internal scour surfaces. Rooted, muddy sandstones with paleosols cap the reconstructed sequence.

The low degree of crossbed variance, both in individual channel-form sandstones and between superimposed channels, is used by Hubert and Forlenza (1987) as a preliminary indication that despite the fining-upward grain sizes, these were low-sinuosity fluvial systems. The authors then use the predominance of trough crossbeds to infer the presence of sinuous-crested dunes in the channels; in contrast, lingoid bars are reconstructed from sedimentary packages consisting of crossbeds overlain by horizontal-planar bedding. The modal cycles containing these structures are thought to represent sinuous-crested dunes in the deeper channels, and the dunes are thought to have decreased in height as the channels filled with sediment and migrated. The horizontal-planar bedding common in the upper part of the modal sequence is interpreted as representing interchannel sand flats. Hubert and Forlenza suggest that these sandy to conglomeratic deposits resulted from braided rivers that carried a sand and gravel bed-load, draining the basin and flowing down its axis. Vegetated overbank and mid-channel islands are recorded by the rooted muddy sandstones.

This assemblage of sandstones and conglomerates grades upward into a finer-grained sandstone sequence in the upper parts of the formation, reflecting the erosional destruction of source-area relief through time. Simultaneously, the lithofacies change to finer-grained and thinner modal cycles ($\frac{1}{2}$ to $4\frac{1}{2}$ meters thick), displaying a reconstructed vertical sequence of sedimentary structures similar to that of the coarse-grained lithofacies.

Although the overall grain-size fines upward through individual sandstone units, Hubert and Forlenza use the rarity/absence of rippled deposits, lateral accretion surfaces, and fine-grained lithologies—plus the indications of wide, shallow channels and the great lateral continuity of crossbed cosets—to rule out high-sinuosity environments and to infer in their place a braided fluvial system with sandy bed-loads. Braided channels with migrating dunes and lingoid bars are thus the inferred depositional environments. In addition, low-water erosional scours are noted within the bar facies, while the sand-flat facies consists of a "nonsystematic mixture" of trough and planar crossbeds. Hubert and Forlenza suggest the sand flats within the channel complex of the South Saskatchewan River as a possible modern analogue.

All of the braided sequences discussed so far contain sedimentary structures of a size commensurate with the basin and the inferred catchment areas of the fluvial systems. In constrast, Thompson (1970a) and Steel and Thompson (1983) have described an unconventional braided sequence of Triassic conglomerates in central England that contains very large gravel crossbeds (up to 4 meters thick) and beds of horizontally bedded gravels that are up to 20 meters thick. Although these deposits are also located in a fault-bounded rift basin, the paleogeographic reconstructions (Fitch, Miller, and Thompson 1966; Audley-Charles 1970; Holloway 1985) indicate that the rift basins of central England provided passageways for more broadly regional drainage, from source areas as far away as northwestern France, through England to the area of the Irish Sea.

Steel and Thompson (1983) interpreted the coarse conglomerate deposits as being the product of an exceptionally large-scale braided fluvial system for which no modern size analogue exists. By extrapolating from the smaller sedimentary structure of present gravelly braided systems, Steel and Thompson reconstructed a system of "medial or midchannel bars with a two-tier structure (subaqueous and partly emergent portions)," which they used to explain upward-coarsening sequences where horizontally bedded (bar-head) conglomerates overlie cross-stratified (bar-slipface) conglomerates. The associated thick, horizontally bedded conglomerates were inferred to be longitudinal bar deposits. The authors also suggested that the stages of discharge and the position a deposit occupied within a forming bar were the factors controlling the degree of sorting and the amount of matrix support of clasts in the deposit.

Although longitudinal regional drainage has not been inferred through the rift basins that contain very coarse Late Permian–Late Triassic conglomerates of central Spain, the scale of structures and many of the sedimentary characteristics of these rocks are similar to those of the British conglomerates. Ramos and Sopeña (1983) and Ramos, Sopeña, and Perez-Arlucea (1986) have isolated several different fluvial lithofacies from the complex conglomerates in this rift system. The most common facies consists of widespread, massive conglomerate sheets ½ to 1½ meters thick and tens of meters in lateral extent. The rocks are clast supported, and the 5- to 15-centimeter clasts are often imbricated and arranged in crude horizontal beds of alternating clast sizes. The meter-scale conglomerate sheets are commonly amalgamated, separated by extensive scour surfaces.

Ramos and Sopeña suggest that these conglomerates formed as complex longitudinal bars in braided-fluvial systems. The beds of alternating clast sizes are interpreted as the record of fluctuating discharge; thus coarse, open-framework gravels indicative of high discharge commonly underlie finer gravels (with infiltrated sandstone) deposited during stages of waning flow.

No evidence for downstream bar slipfaces is reported, but the horizontally bedded conglomerates commonly occur adjacent to filled channel forms and/or distinctive, diagonally bedded conglomerates. Since imbrication of the clasts in the horizontally bedded facies is oblique to the strike of the diagonal bedding, Ramos and Sopeña (1983) infer that the diagonal beds were deposited in the channels as lateral accretion modifications to the main bars during channel migration and waning discharge.

Thus, models of braided-river deposits based on modern fluvial systems have been useful in the interpretation of many of the ancient braided fluvial deposits, but much remains that is ambiguous in these sequences because of their complexity and wide range of variability. Given that the recognized limits of variability in modern fluvial systems continue to expand, this is not surprising. Still, a geologist must make attempts at interpretations, using the most appropriate models available. Without modern models (as was shown in Part I of this book), geologists are allowed almost unlimited speculation. Much of the refinement possible within the science of sedimentology may lie in knowing how much interpretive detail the data can support, as well as in knowing which data to record in order to construct reasonable interpretations.

Meandering-river Deposits

Although meandering rivers are still being studied and different models for their deposits are still being generated, the basic process of lateral accretion common to meandering-river systems is well understood. This process often allows preservation of complete vertical channel-bottom-to-overbank sequences, especially in rapidly subsiding basins such as rift basins. Vertical sequences of grain sizes and sedimentary structures within these deposits are repetitive and easily recognized, once the norm for a formation is recognized. Expressed as cycles, as cyclothems, as sequences, or in accordance with whatever terminology is currently in vogue, these fining-upward patterns have been recognized in at least five of the rift basins discussed here.

Such patterns occur in sandstones in the Pomperaug outlier, interbedded with braided-river deposits, as a distal facies equivalent to the alluvial-fan and braided-river deposits of the New Haven Arkose in the Hartford basin (Hubert et al. 1978). Because they were distal equivalents to braided facies, they locally onlapped and now overlie the coarser facies in basins where inactive tectonics allowed the erosional lowering of source-area relief. Similar examples of onlap have been described by Brown (1980), Steel (1974), and Clemmensen (1980), in Morocco, Scotland, and Greenland, respectively.

Few of the fining-upward sequences noted in the rift basins have been adequately described. Surprisingly, Henson's (1970) brief description of such trends in the Triassic of South Devon, England, is probably the most complete (or at least best conforms to the standard model characteristics). Henson inferred "channel and point bars" in a high-sinuosity fluvial system, based on 1.25- to 3-meter-thick sedimentation units that consist of a lower scour surface, basal sandstones with "large" crossbeds, local rip-up clasts, and horizontal planar bedding. The sequence grades up into finger sandstones with smaller crossbeds, and is capped by silty claystones. The sedimentation units are locally superimposed to create multistory sandstones that are up to 15 meters thick.

Lateral accretion surfaces are occasionally visible in such fining-upward

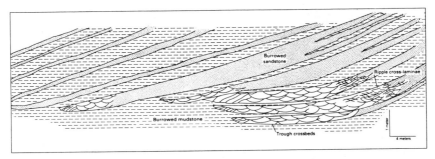

Figure 6-3. Schematic diagram of inferred lateral accretion surfaces from the Durham basin. Each sandstone unit is coarse-grained and contains mudstone rip-up clasts in the basal crossbedded zones, but each becomes fine- to medium-grained in the burrowed/massive upper zones. Note the vertical exaggeration. (From Smoot 1985, Figure 2.1. Reprinted with the permission of the U.S. Geological Survey)

deposits (Smoot 1985; Clemmensen 1980). Smoot (1985) has described point-bar beds that dip about 5 degrees and are about 40 meters wide (figure 6-3); within these point bars, beds of mudstone were deposited between successive sandstone lateral-accretion beds. Point bars of this type may record rapid lateral migration or perhaps high variability in discharge, but the significance of different types of point-bar bedding is poorly understood at present.

The mudstone-to-sandstone ratio usually increases in the formation as a whole in association with these deposits, and the channel deposits are finer-grained than are the correlative alluvial-fan/braided-river deposits. In Morocco, where extensive, relatively well-exposed outcrops have been studied on a broad scale, Brown (1980, p. 999) has described the alluvial architecture of these deposits as "upward-fining coarse-grained sandstone units [that] form elongate belts set *en echelon* within intervals of interbedded mudstone, siltstone, and fine-grained sandstone. These are deposits of highly sinuous meandering river channels and adjacent flood plains" (figure 6-4).

Detailed, structure-by-structure measured sections and outcrop descriptions of deposits left by meandering fluvial systems in the rift basins are rare, but it is apparent from existing descriptions that geologists feel comfortable with the standard interpretations of these sediments and with the application of meandering models in general. At present, considerably less effort is being devoted to them than to the less-well-known braided deposits. Either geologists are on the right track and everything is fitting nicely, or meandering-

Figure 6-4. High-sinuosity fluvial deposits, Essaouira basin, Morocco. Note the high ratio of (red) fine-grained overbank material to (white) channel sandstones. (Photo courtesy of R. H. Brown)

channel sediments form too small a percentage of the formations to be worth much concentrated effort. A less pleasant alternative is that we have become complacent with a common-knowledge model, continuing a scientific tradition that once found expression in the offhand application of the estuarine model to the Hartford basin (see figure 6-5).

Overbank Deposits and Postdepositional Modifications

Accumulations of overbank material in the rift basins and elsewhere have been given little attention, in large part because they are usually poorly exposed. Such sediments are subject to long periods of nondeposition and subaerial exposure, during which they undergo modification by both organic and inorganic processes. Few of the overbank deposits that have been studied are in their original form, and thus the diagnostic characteristics of overbank strata are commonly those produced by modification processes rather than those produced by depositional processes. These deposits are often termed *floodplain* deposits, although other authors use that term to include channel sandstones and exclude alluvial-fan mudstones.

Figure 6-5. Two examples of diagonal bedding planes within sandstone deposits, Hartford basin, Connecticut. Structures of this type have been attributed to point bars in high-sinuosity fluvial systems. The sandstones in neither example, however, contain typical basal scours and rip-ups; nor do they exhibit crossbedding that grades up into finer structures and grain sizes. Most of the structures consist of planar bedding. These sandstones need further study. (Both photos by the author)

Overbank environments are most often recognized only as the mudstones or sandy mudstones at the top of a fluvial channel sequence, although details of primary depositional characteristics of overbank deposits have been described briefly by Hubert, Reed, and Carey (1976), LeTourneau (1985a, 1985c), and Gore (1985a). LeTourneau has described horizontally laminated red siltstones and silty mudstones that contain climbing ripple stratification and (locally) lenses of crossbedded sandstone. The sandstones are interpreted as ephemeral channels on the floodplain, but the siltstones and mudstones are not discussed beyond their characterization as "flood-plain" deposits.

Gore (1986) has described floodplain deposits from the Culpeper basin, interpreting half-meter-thick, lenticular, coarsening-upward siltstone-to-sandstone beds as "levee or crevasse splay deposits." Gore has also reported a 2-centimeter-thick "ostracodal limestone" that occurs near the top of a fining-upward siltstone-mudstone sequence and, in conjunction with fossil evidence, is interpreted as a "floodplain lacustrine deposit."

Hubert, Reed, and Carey (1976) described overbank deposits as horizontally laminated mudstones up to 4 meters thick, locally containing carbona-ceous shales or thin graded beds. The graded beds were inferred to represent deposition of suspended sediment by settling in flood-inundated over-bank areas.

Soon after deposition, these fine-grained deposits were subject to organic modification processes—the traces of which include root casts, vertebrate tracks (figure 6-6), various animal trails, and invertebrate burrows (figure 6-7). The most commonly identified burrow trace, an ornamented tubular burrow with a meniscate filling, is that of *Scoyenia*. This trace fossil is most commonly interpreted as a crayfish burrow and is usually taken as evidence for generalized "floodplain" deposition in North America (Stevens and Hubert 1980; Hubert, Gilchrist, and Reed 1982), on the basis of its apparent association with fossilized crayfish parts in the Durham basin of North Carolina (Olsen 1977). Turner-Peterson (1980), however, inferred that *Scoyenia* occurred in near-shore lacustrine deposits in the Newark basin; and in the rift-basin deposits of Greenland, Bromley and Asgaard (1979) have recog-nized a refined *Scoyenia* assemblage of trace fossils, suggesting that the assemblage indicates "extremely shallow lacustrine conditions." They also inferred that *Scoyenia* itself is the trace of an insect. Thus, while *Scoyenia* may be indicative of a generalized floodplain environment, the specific subenvironment or subenvironments and the animal that made them are still in question.

Inorganic processes also modified floodplain sediments and the abandoned,

Figure 6-6. Vertebrate track in overbank mudstones, Central-N'Fis basin, Morocco. Note the preserved, fine texture of the skin and claw traces. (Collected by P. Biron; photo courtesy of J. Beauchamp)

sandy, channel-fill sequences. These modifications include load casts of sandstone into the soft, muddy sediments (figure 6-8), desiccation cracks (often multiple episodes) (figure 6-9), and the features of soil-forming processes. Caliche profiles (figure 6-10) are the most common soil types found in most of the rift basins. The most thoroughly studied of these is found in the New Haven Arkose of the Hartford basin (Hubert 1977, 1978), where caliche has been convincingly documented and used as an indicator of a seasonal but semiarid climate.

Caliche-type carbonates often form as isolated nodules or kankar within muddy floodplain soils (figure 6-11). When the floodplains are eroded by fluvial processes, the muddy sediments wash away and the larger, more resistant kankar nodules are incorporated into the fluvial deposits as channel-base intraclasts. As a result, carbonate nodules are commonly found at the base of channels, even when little or nothing remains of the contemporaneous overbank mudstones and soils.

An indication of the resistance to erosion of the cohesive (and often rooted) mudstones is the steepness of some of the erosional channel margins

Figure 6-7. Organic modification of overbank deposits, Hartford basin, Connecticut. (A) Burrowing in overbank sandstones, and bioturbated, structureless texture in overlying mudstones. (B) Bedding plane showing tracks and trails on the plane and penetrations by vertical burrows. (Both photos by the author)

Figure 6-8. Gravity-loading structure: fluvial sandstone introduced into an area of unconsolidated muddy overbank sediments, Hartford basin, Connecticut. (Photo by the author)

(figure 6-12). The development of calcareous soil profiles in these deposits enhanced the mechanical resistance to erosion inherent in mudstone and claystone deposits. Channel margins cut into sandy deposits are rarely as steep as these.

Recently, Smoot and Olsen (1985) have presented a preliminary report of a study of the neglected overbank deposits. They have recognized four major divisions of mudstone types, each of them seemingly massive at first glance but with subtle differences in textures and associated features. Smoot and Olsen suggest that two of these mudstone types represent floodplain environments. The first, interpreted as recording vegetated floodplains, is characterized by rooted, calichefied, and mudcracked massive mudstones containing *Scoyenia* trace fossils. The second—which evidently records a variety of floodplain deposits, "including shallow margins of lakes, delta plains, and fluvial floodplain-overbank settings" (p. 30)—is also characterized by *Scoyenia*, but includes vertebrate tracks, ostracods and other aquatic fossils, and remnant inorganic structures such as microlaminated mudstones, climbing ripples, and soft sediment deformation features, all within the overall massive mudstone setting. The reported features have been simplified here, and the

Figure 6-9. (A) Desiccation cracks infilled with caliche carbonates. Gribun, northwest Scotland. (Photo courtesy of B. J. Bluck) (B) Multiple episodes of development of desiccation cracks in fine-grained fluvial sandstones. Kerrouchen basin, Morocco. (Photo by the author)

recognized complexity of these deposits promises to increase. This type of study has good potential for enhancing the knowledge of overbank depositional environments.

Facies: Patterns and Controls

The major vertical patterns in fluvial sequences are similar to those in alluvial-fan deposits. Braided sequences commonly grade into meandering sequences distally and upsection, accompanied by a general decrease in grain size from conglomerates to sandstones. This pattern has been attributed either to change in climate (Steel 1974) or to decrease in relief of the source area during tectonic stability (Hubert et al. 1983; Steel and Wilson 1975; Clemmensen 1980).

Sequences that coarsen upward also occur in rift basins, where braided systems followed finer-grained meandering ones. Steel and Thompson (1983), and Ramos, Sopeña, and Perez-Arlucea (1986) attribute such coarsening-upward patterns in the basal parts of the basin fills in North Staffordshire, England, and in central Spain to the initial tectonism associated with basin formation. Clemmensen (1980) described a similar coarsening-upward sequence in Greenland but concluded that insufficient data were available to make a choice between tectonic or climatic causes.

Most rift-basin fluvial deposits interfinger with alluvial fans upslope but are interbedded with a variety of deposits downslope. The downslope environments varied with the basin setting. In blocked or interior-drainage situations, especially during climatic trends toward increased precipitation, lacustrine or paludal sediments interfingered with fluvial deposits. During episodes of through-drainage, fluvial and overbank deposits extended throughout the central portion of the basins. At other times (probably under arid, closed-basin conditions), playa environments were found directly at the edge of alluvial fans and their associated fluvial deposits, and these two facies became interbedded.

After initially divergent radial paleodrainage across the conical fan surfaces, rift-basin fluvial systems were often diverted to flow longitudinally down the axis of the basins. Exceptions to this pattern occurred in central England and in the High Atlas region of Morocco, where paleodrainage was directed down the axis of the basins at all times, and in the Hartford basin, where paleodrainage seems to have usually been oblique across the basin (see figure 4-7) during fluvial deposition (Hubert et al. 1978).

Two factors were capable of causing further reorientations of the local paleodrainage. In the Fundy basin, the growth of alluvial fans into the basin may have deflected axial drainage, as well as offsetting it laterally away from the faulted margin from which the fan was growing (Nadon and Middleton 1985). The second factor is illustrated in western Morocco, where tilted surfaces of synsedimentary fault blocks within the basin diverted local paleodrainage. Fluvial systems that flowed across the tilted surfaces were deflected laterally across part of the basin into the adjacent internal grabens, where they resumed their longitudinal patterns (Brown 1980).

Internal fault blocks may also occur within the Durham basin of North Carolina (Bain and Harvey 1977, noting aeromagnetic evidence), but paleocurrent data from the associated sediments have not been published. A plot of the regional crossbedding vectors of these sediments would help determine whether paleodrainage was affected in a manner similar to that shown by Brown's study. Leith and Custer (1968) did note local variations

Figure 6-10. Variability in development of caliche soil profiles. (A) Caliche profile consisting predominantly of densely packed, calcified root casts in red floodplain mudstones, below fluvial sandstone (hammer for scale), Hartford basin, Connecticut. (Photo courtesy of J. F. Hubert, from Hubert 1978, Figure 6. Reprinted with the permission of *Palaeogeography, Palaeoclimatology, and Palaeoecology*) (B) Thick zone of well-developed caliche pillars (hammer for scale), Inch Kenneth, northwest Scotland. (Photo courtesy of B. J. Bluck) (C) Close-up of expanded-rock fabric typical of some caliches—a lattice of carbonate in mudstone (lens cap for scale), Hartford basin, Connecticut. (Photo by the author)

Figure 6-11. Authigenic carbonate kankar nodules (A) disseminated in floodplain mudstones, and (B) secondarily incorporated into an adjacent channel-base intraclast conglomerate, Hartford basin, Connecticut. (Both photos by the author)

Figure 6-12. Abrupt terminations of fluvial channel margins against floodplain mudstones, indicating resistance of the mudstones to erosion. (A) Fine- to medium-grained sandstone, Pomperaug outlier, Connecticut. Note the slight overhang of cohesive mudstones over undercutting channel sandstone. (B) Coarse-grained sandstone, Hartford basin, Connecticut. (Both photos by the author)

in paleodrainage in the Durham basin, but they attributed it to probable stream meandering.

Summary

Three depositional environments were covered in this chapter. First, braided-stream deposits were considered. They can be correlated well with the most generalized models of braided fluvial systems but poorly with existing stream-specific modern analogues. Second, the generalized meandering-stream model was shown to have application to some of the fluvial deposits, but as yet little effort has been directed toward more specific interpretations of these rocks. Third, the fine-grained overbank deposits were mentioned. These deposits are just beginning to be studied carefully.

The character of the fluvial deposits in the rift basins, as well as the concurrent paleogeography, seem to be controlled primarily by the local tectonics, by rates and frequency of uplift and subsidence, and by basin size and shape. Climate is a secondary control. No attempt has yet been made to correlate regional tectonic or climatic events, as recorded by fluvial sedimentation patterns, between basins on a continental scale.

Data Collection and Interpretation Techniques Illustrated by Studies of Rift-Basin Fluvial Deposits

Recognition of primary sedimentary structures and their hydrodynamic regimes, to quote the title of the landmark 1965 SEPM publication, is the common starting point for outcrop examination and subsequent interpretation of fluvial deposits. Once the primary structures are understood, their associations can be noted and, more importantly, their distributions and relative orientations can be mapped. For instance, numerous authors have noted horizontal-planar bedding associated with tabular crossbedding and have inferred from this association backbar and bar-slipface microenvironments, respectively; thus, they have interpreted the sediments as bar deposits. The relative positions of these structures have rarely been reported, however, and Miall's (1985) architectural-element analysis technique has not yet been widely applied.

However, Bluck (1967), Hubert et al. (1978), Ramos, Sopeña, and Perez-Arlucea (1986), and Steel (1974) have published sketches of outcrops in different basins that indicate the complexity of the arrangements of sedimentary structures. Such figures indicate the positions of structures and the orientations of directional features relative to the outcrop, so that an idea may be formed of the position of the crossbeds with respect to paleoflow. The variability of directional features within an outcrop, between different beds, or even within a single extensive sedimentary structure, provides information on paleoflow variability within the fluvial system and often gives an indication of the shape of the bed form.

Although they have not been reported from the rift basins, bed-form morphologies may also be inferred from their patterns on infrequent bedding-plane exposures. Field geologists should note and use this unique type of data. An example of uncommon data from strata in a rift basin is percussion marks on the larger clasts, indicating "repeated collisions during traction transport" (Nadon and Middleton 1985, p. 1193). There is rarely a single golden key to environmental interpretations, however, and a geologist had best be prepared to note and use the commonly occurring assemblages of sedimentary structures observable in standard vertical outcrops.

Continuing with the building-block approach, one may note the vertical trends of grain sizes and sedimentary structures within sandstone beds (figure 6-13). In the deposits of braided streams, a significant data base and

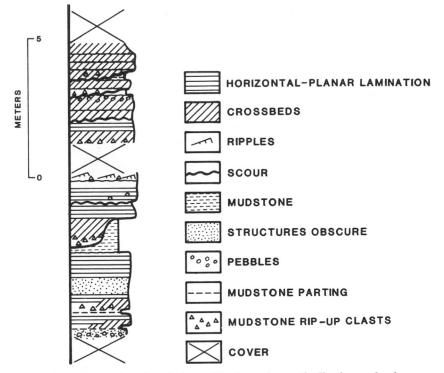

Figure 6-13. Example of "vertical" profile through steeply dipping rocks along the Pomperaug River, Connecticut. Despite the absence of lateral exposure, the vertical sequence of sedimentary structures can be used to infer a braided fluvial system.

statistical methods may be needed to recognize repeated or probable vertical sequences. As discussed, the vertical patterns of meandering-stream deposits are commonly more obvious, unless slowed rates of deposition have allowed the development of a meander belt where numerous sweeps of the laterally migrating channel have resulted in reworking of sediments and a multistory sandstone deposit.

Where data permit, the alluvial architecture of a basin-fill sequence provides information on the depositional environment. Such extensive outcrops as Brown's (figure 6-4), showing meander-belt sandstones isolated within a fine-grained matrix, can be contrasted to the sandier sequences of braided deposits (figure 6-2). Several authors (Clemmensen 1980; Stevens and Hubert 1980) have cited high sandstone-to-mudstone ratios as evidence for braided systems, but there is as yet no widely accepted cutoff ratio that defines the boundary between meandering and braided; ancient braided systems with high percentages of overbank mudstone may, in fact, have existed where tectonic damming trapped both coarse and fine sediments in a basin despite the presence of significant paleoslopes (Bhattacharyya and Lorenz 1983). If the sandstone-to-mudstone ratio is to be used, the best approach is probably to compare deposits within the basin with each other, rather than with an outside standard. This type of comparative approach factors out many of the possible variables.

On a basin-wide scale, regional fluvial crossbedding vectors have proved to be of great value in paleogeographic reconstructions (Hubert et al. 1978). They may indicate both regional patterns and more subtle structural variations within the basin (Brown 1980). Such studies have not yet been undertaken or published for many rift basins.

7 / **Paludal Deposits**

Coals constitute the most geographically and temporally restricted sedimentary deposits in the rift basins. Except for a Late Triassic–Early Jurassic sequence in eastern Greenland mentioned by Clemmensen (1980), all of the known paludal deposits in this rift system are found in the basins of Virginia and North Carolina (figure 7-1) and all are of early Late Triassic (middle Carnian) age (Robbins, Wilkes, and Textoris 1987). They are among the oldest dated deposits in the North American rift basins. Robbins, Wilkes, and Textoris (1987) have noted that the middle Carnian was an age in which coal was forming in a number of areas worldwide, although no coals are known from the equivalent strata in England, Morocco, or Greenland.

Coal was mined from the Triassic paludal deposits in the 1800s and early 1900s, and they were the subject of a number of early descriptive economic studies. Since then, most studies of the basins have mentioned the coals only briefly, and detailed sedimentological studies have not been published. Robbins (1983, 1985) and Robbins, Wilkes, and Textoris (1987) have recently studied the coals from the viewpoint of regional basin settings and analogous modern processes, in order to understand the interactions among tectonic environment, depositional environment, and resulting economic deposits. Details of outcrop-specific environmental interpretations have not yet been published, but Ressetar and Taylor (1987) have a manuscript in preparation that will fill some of this vacuum.

In this chapter, some pertinent descriptions from the older literature are integrated with the current concepts of paludal depositional models. The available data are insufficient in most cases for specific reconstruction of subenvironments such as the depositional models of raised, floating, or low-lying swamps listed by McCabe (1984), but the stratigraphic position and

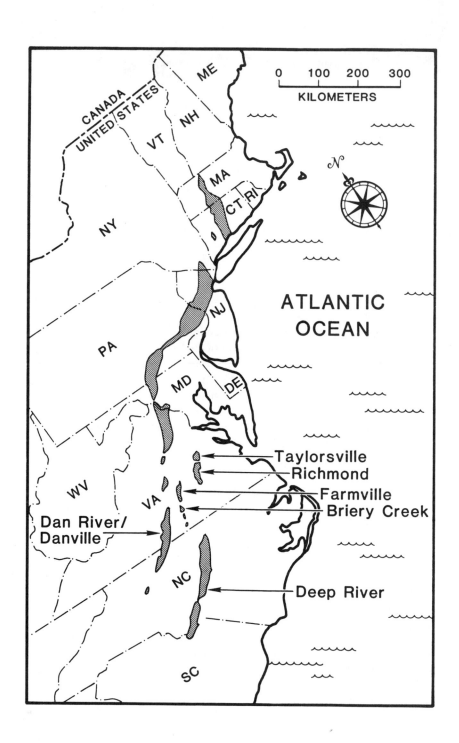

CANADA
UNITED STATES

ME
NH
VT
MA
CT RI
NY
NJ
PA
MD
DE
WV
VA
NC
SC

ATLANTIC
OCEAN

0   100   200   300
KILOMETERS

N

Taylorsville
Richmond
Farmville
Briery Creek

Dan River/
Danville

Deep River

lithologic associations of the coals often allow inferences of probable generalized depositional settings. Study of the Triassic coals is hampered by their
poor exposure (figure 7-2), yet because of numerous exploratory drillholes,
the coals are one of the few lithologies that have been delineated in three
dimensions. These Triassic coals have closer affinities to the nonmarine coals
of the numerous Chinese Mesozoic basins (which western geologists are just
becoming aware of) than they have to the Carboniferous coals of eastern
North America or to the Cretaceous coals of the Rocky Mountains region.

Coals in general—including the Triassic coals—have often been used as
evidence for tropical paleoclimates of high precipitation or high humidity.
As pointed out by McCabe (1984, p. 22), however, modern peats are forming
in a variety of climates at a variety of latitudes, and "coal itself is not a good
paleoclimatic indicator." It has also been suggested that coals do not form
where clastic sedimentation is rapid, but recently this concept has been
questioned. Flores and Warwick (1984), and Sitian et al. (1984) have even
documented thick coal deposits that formed in other intermontaine basins
at the toes of alluvial fans, where ground water discharge from the fans
apparently kept water tables high enough for prolific plant growth and at the
same time maintained the stagnant water conditions necessary to retard
oxidation and decay of dead plant material.

Thus, the question may not be why coals formed in some of the rift basins,
but why coals did not form in more of them. It should be noted in this
regard, however, that many of the rift-basin coals are more closely associated
with lacustrine strata than with alluvial-fan conglomerates. In this respect,
Robbins (1983, p. 652) has suggested that "highly alkaline [modern] lakes do
not seem to preserve plant tissues, so coal was not found associated with the
[ancient] saline lake bed deposits of the northernmost (Triassic) basins. . . ."
Their absence may also have been due to a shift toward truly arid conditions—
either northward, as indicated by the eolian deposits of Fundy and Greenland
(climatic trend suggested by Hubert and Mertz 1980), or with time, since the
coal-bearing deposits are the oldest of the known sediments in the rift-basin
fills of North America. In either case, the Late Triassic–Early Jurassic coals of
Greenland do not fit the proposed patterns.

Figure 7-1 *(facing page)*. **Coal deposits in the Triassic–Jurassic rift basins are
limited to the basins named in this figure, in the mid-Atlantic states of the
United States (Virginia and North Carolina), as well as to the strata of early Late
Triassic age.**

Figure 7-2. Coal outcrop in the Farmville basin, Virginia. Three coal seams are exposed in this road cut, but neither the coals nor the associated fine-grained sedimentary rocks produce exposures amenable to extensive sedimentological study. (Photo courtesy of E. I. Robbins)

Coal Deposits of the Deep River Basin

Reinemund (1955) has provided the most detailed published information to date for the interpretation of coals in a rift basin. His study concerned coals of the Cumnock Formation in the Sanford subbasin (the southern half of the Deep River basin; see figure 7-1). The coals in this basin occur as part of a tectonically influenced trend in the basin fill—the coals overlying a thick fining-upward fan–fluvial–lacustrine sequence, and underlying a series of mudstone deposits (probably lacustrine) that continue the fining-upward trend (figure 7-3). The coals occur at two horizons in the lower third of the overall fine-grained Cumnock Formation. Both of the coaly intervals are thickest in the same general area (figure 7-4), which coincides with the thickest portion of the Cumnock Formation. This thickening occurs toward the unfaulted edge of the basin, suggesting that the Cumnock sediments (including the coals) filled in preexisting topography of the alluvial paleoslope.

The coals grade into coarse-grained clastics (probably alluvial-fan deposits) southeastward toward the major border fault, and longitudinally down

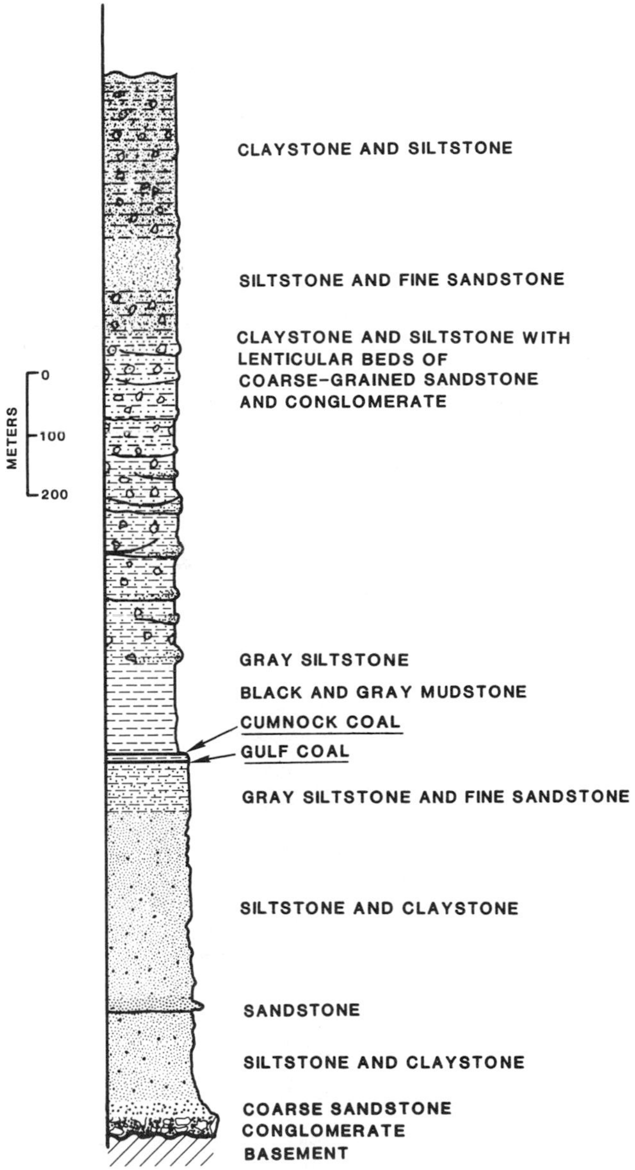

Figure 7-3. Generalized lithologic column of the rocks in the Deep River basin, showing that the coal deposits are found near the top of a formation-scale fining-upward sequence, in close association with the unoxidized fine-grained lacustrine facies of the Cumnock Formation. (After Reinemund 1955, Figure 6. Reproduced with the permission of the U.S. Geological Survey)

the basin axis to the southwest. Thus, the coals are localized in the northwestern part of the Sanford subbasin. Traces of coal also apparently occur in the southern part of the Durham subbasin adjoining to the north. The localization of the coal deposits points to the importance of the local tectonic controls on coal deposition in these rift basins. A purely climatic origin presumably would have produced more widespread coal deposits.

Reinemund's isopach maps of the coal (figure 7-5) were made primarily

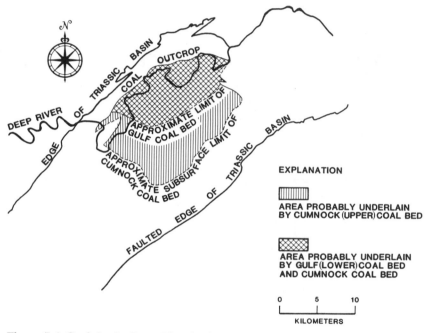

Figure 7-4. Coals in the Deep River basin and most other rift basins are restricted to small areas within the basins, suggesting that they formed in localities that were temporarily free from clastic influx as a result of heterogeneous tectonism. (After Reinemund 1955, Figure 38. Reproduced with the permission of the U.S. Geological Survey)

Figure 7-5 *(facing page).* Isopach maps of coals of the three coaly zones of the Cumnock Formation, Deep River basin. (A) Top bench. Note that coal and sandstone deposition were mutually exclusive at this horizon. (B) Main coal. Note that, as mapped, coal distribution does not reflect the present-day faults. (Cross section A–A' is shown in figure 7-6.) (C) Lower benches. All coals below the main coal are combined in this map to portray net combined coal thickness, reflecting the mapper's economic interest in the coals, rather than sedimentological interest. (After Reinemund 1955, Plate 7. Reproduced with the permission of the U.S. Geological Survey)

for economic purposes, but they lend themselves to some generalized sedimentological interpretations. The various beds of the Cumnock coal show patterns of thinning and thickening transverse to the basin axis— patterns that also seem to be reflected in subsequent depositional events. The thickest beds of the lowest group of coals occur primarily in areas where

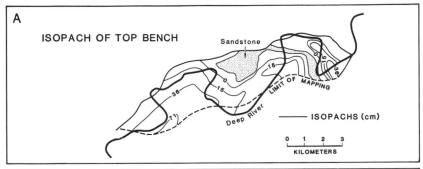

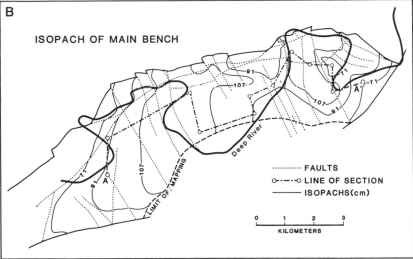

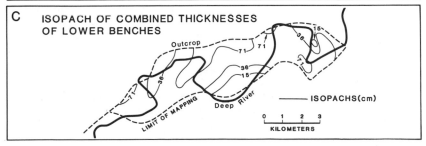

the succeeding main coal bed is thin. These thin areas of the main bed were the locus of deposition of a wedge of sandstone, described as having a maximum thickness of 11.3 meters (37 feet) and as "thin[ning] rapidly in all directions" (Reinemund 1955, p. 33). The deposition of sandstone apparently overwhelmed the swamp environment; thus the youngest bed of coal is restricted to areas farthest from this clastic depocenter. The general patterns of thinning and thickening, however, are not indicative of any specific coal-forming environment.

Syndepositional faulting within the basin may have caused the local thinning and thickening of individual beds, although (as mapped by Reinemund) the lateral variability in coal thicknesses does not seem to reflect the known fault patterns. If the variation in thickness is a primary sedimentary feature, it does not reflect a uniform-thickness or concentric pattern that might be associated with the filling of a lake by a peat swamp. Most of the coals are relatively discontinuous and difficult to correlate (figure 7-6). Only the main coal bed of the Cumnock has consistency in extent or thickness, and the presumed consistency rests on the assumption that the thickest coal in any borehole is in fact correlative with the thickest coal in the adjacent borehole, in order to make the basin-wide correlation shown. Some caution should be exercised when correlating beds that are less than a meter thick across distances of kilometers. In sum, much work remains to be done in defining the depositional environments of these coals.

There was probably a fine balance between basin subsidence, clastic sedimentation, and peat formation during this time. Yet the influx of 11

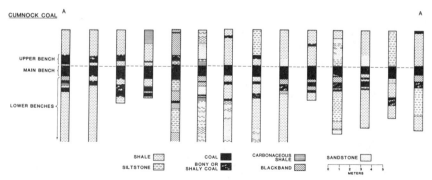

Figure 7-6. Cross section between drillholes, showing lateral variability of coal beds, Cumnock Formation, Deep River basin. Only the main bench is correlatable with any degree of certainty, and even this assumes that the thickest coal in any hole is the main bench. See figure 7-5 for the location of this section. (After Reinemund 1955, Plate 8. Reproduced with the permission of the U.S. Geological Survey)

meters of sand at one edge of the area merely localized peat deposition, rather than filling all of the coal swamps with coarse clastic deposits. On the basis of the close association of the Cumnock coals with lacustrine shales, Robbins, Wilkes, and Textoris (1987) suggested that the coals formed in swamps that fringed the rift-basin lakes and later (as they developed) extended into and filled the lakes.

The episode of coal deposition in the Sanford subbasin was succeeded by lacustrine conditions, which were terminated when renewed tectonic activity reintroduced coarse clastic material into this distal part of the basin.

Comparison with Other Triassic Rift-Basin Coals

Coal deposits in the other rift basins of Virginia and North Carolina also occur above deposits of an earlier phase of tectonic activity, associated with the finer-grained deposits of waning or stable tectonics. In the Dan River basin of North Carolina, coals occur at the top of an initial basin-fill sequence (Thayer 1970). They are associated with lacustrine shales and lie within the lower third of the fine-grained Cow Branch Formation.

Thayer (1970) listed some of the details of this coal, noting that it occurs only in thin, lenticular layers. He described autochthonous plant fragments (indicating low energy) and the auxiliary minerals vivianite (an iron phosphate), pyrite, and siderite concretions, indicative of subaqueous environments. The association with lacustrine shales suggested to Thayer that the coal accumulated "in swamps and deltas along the lake margins" (p. 16). Thayer also noted a number of features in the associated Cow Branch lake deposits that imply arid conditions, including desiccation cracks, raindrop impressions, and casts of halite crystals. Although these characteristics do not fit readily into standard models of paludal environments, the coals in this particular formation reach 3 meters in thickness. It would be useful to know the vertical and lateral distributions of coal thicknesses compared with the distribution of perennial and ephemeral lake characteristics, but such data have not been published.

Coals in the Farmville, Taylorsville, and Richmond basins also occur near the top of an initially conglomeratic, fining-upward sequence. A study by Ressetar and Taylor (1987) suggests that the coals in the Taylorsville basin were deposited during a stage of tectonic stability, in an environment of poorly drained floodplains crossed by meandering streams. Crossbedded sandstones (meandering channels) are associated with sequences of interbedded dark mudstones and horizontal-planar bedded sandstones (figure 7-7) that are inferred to be the result of flooding in overbank areas. The thinness of the black mudstones and the presence of associated nonlacustrine deposits

Figure 7-7. Thin horizontal-parallel bedded sandstones within black mudstone sequence in the Taylorsville basin. Such sandstones, locally graded or containing crossbedding, are interpreted as flood deposits (derived from local fluvial channels) into floodplain lakes in which local coals were being deposited. Scale is 15 centimeters long. (Photo and interpretation courtesy of R. Ressetar and G. Taylor)

are taken to imply that the interbedded coals were not associated with a major lacustrine environment, but rather were deposited in or along numerous small ponds and oxbow lakes on the floodplain.

Goodwin et al. (1985) have reported rooted underclays beneath some of the coal seams in the Richmond basin, indicating that these coals are autochthonous. In places, coals can be seen to grade laterally into sandstones and shales in this basin, and Ressetar and Taylor (1987) have inferred that the coals were deposited in swamps traversed by meandering, episodically flooded streams.

At present, some of the coals in the Richmond basin are in fault contact with basin-margin granites, and the facies relationships in the coal-bearing strata are obscured by faulting (figure 7-8). The apparently rapid lateral transitions between the sandstones and other lithologies are entirely due to postdepositional faulting, rather than to facies changes, although within the fine-grained deposits, coals do grade rapidly into carbonaceous shales.

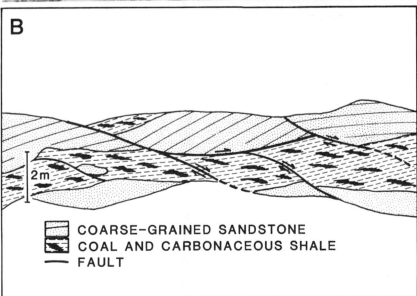

Figure 7-8. Faulted coal-bearing deposits in the Richmond basin. (A) Photo. (B) Tracing from photo. (Photo and interpretations courtesy of R. Ressetar and G. Taylor)

Wilkes (1982) has diagramed coals in the Farmville basin as distally equivalent to the border-fault alluvial fans. Coals occur in the southern end of this basin and near the unfaulted basin margin, in much the same type of setting as that reported by Reinemund (1955) for coals in the Deep River basin. Crossbedding shows that the paleodrainage was southward down the axis of the basin, and thus toward the coal-swamp environments (Wilkes 1982). The coals seem to have accumulated on the topographic lows within the basins, in areas that were free from clastic sedimentation (possibly because of tectonic stability).

Roberts (1928) noted that vertebrate tracks were noticeably less frequent within the coal-bearing sequences. This presumably has more to do with the poorer preservation potential of tracks in soupy paludal and spongy peat deposits than it has to do with a diminished fauna. Roberts also provided the drillhole lithologic logs from several coal-exploration holes, at least one of which (in the Farmville basin) seems to indicate a repetitive alternation of lithologies, and therefore of environments (figure 7-9). Whether this phenomenon is related to climate, tectonics, or autocyclic processes is unclear.

Facies: Patterns and Controls

Most of the Triassic coal deposits apparently were associated with waning stages of an early tectonic episode and with meandering-fluvial or lacustrine deposits. Both the tectonic history and the fine-grained episodes of deposition may be correlative between the local basins. The coals, however, are not basin-wide deposits; most commonly, they are localized within a specific part of each basin and are restricted to areas remote from the immediate influence of tectonically produced, rapid, alluvial-fan sedimentation.

Robbins, Wilkes, and Textoris (1987) cite a 1975 thesis study of associated megaflora as indicating that coals in the Richmond basin formed in swamps and on floodplains in a tropical, humid climate. There is also evidence for episodes of contemporaneous desiccation and minor evaporite formation in some of the lacustrine shale deposits associated with the coals (Thayer 1970), however, and the two forms of evidence have not been reconciled.

Lateral thinning and thickening patterns of individual coal beds may in part be the result of syndepositional faulting within the basin. The thickness trends do not coincide with known faults, however, and the scale of thickness variation (a few meters) is probably less than the present offset of most faults. The thickness variations more likely reflect local depositional patterns such as proximity to active fluvial channels or rates and amounts of

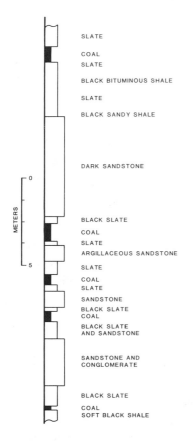

SLATE

COAL
SLATE

BLACK BITUMINOUS SHALE

SLATE

BLACK SANDY SHALE

DARK SANDSTONE

0

METERS

BLACK SLATE

COAL
SLATE
ARGILLACEOUS SANDSTONE

5

SLATE

COAL
SLATE

SANDSTONE

BLACK SLATE
COAL

BLACK SLATE
AND SANDSTONE

SANDSTONE AND
CONGLOMERATE

BLACK SLATE

COAL
SOFT BLACK SHALE

Figure 7-9. Figure drawn from drillhole lithology log presented by Roberts (1928, p. 100) suggests some cyclicity to coal deposits in the Farmville basin, Virginia. NOTE: The fine-grained lithologies (such as "slate" and "shale") have been lumped together.

differential subsidence due to burial compaction of the underlying hetero-geneous lithologies.

Many of the coals are associated with lacustrine deposits, and the cyclic variations through time of the lacustrine environments (discussed in chapter 8) undoubtedly affected the location and extent of the adjacent paludal environments. Narrow lake-margin coal swamps would have migrated across the basin floor as the lakes expanded and contracted; and if the migrations were slow, laterally extensive peat deposits would have been left behind. More extensive and less diachronous paludal environments may be recorded in the Deep River coal beds. Other rift-basin coals are associated with fluvial deposits and probably represent channel-margin swamps and the infilling of floodplain lakes. These coals are more localized, grading laterally into other floodplain deposits over short distances.

Summary

Published evidence is insufficient to sustain a detailed examination of the sedimentology of rift-basin coal deposits for the purpose of assessing existing interpretations or proposing definitive new ones. With increasing knowledge of the associated sand body geometries and sedimentary structures, however, it is probable that the specific models developed from modern rift-basin environments and from the ancient deposits within the interior basins of China will prove useful. In general, the limnic rift-basin coals are usually associated with lacustrine or fine-grained fluvial environments.

Data Collection and Interpretation Techniques Illustrated by Studies of Rift-Basin Paludal Deposits

The literature on the Triassic rift-basin paludal deposits is also inadequate for illustrating most of the different techniques used in the detailed sedimentological study of coal outcrops. Much of the available data come from subsurface economic drilling programs. The use of such data for correlation and isopach studies is common and valid; but before such data can be used for environmental analysis, the average distance of significant horizontal variability should be established. This average distance can be established by means of closely spaced drillholes, and it should be done before conclusions are made based on regional correlations of widely spaced data points.

If cored, drillholes can also provide excellent lithologic and sedimentological detail through one-dimensional, isolated vertical sections. In fact, the core data in these mudstone and coal lithologies are often of better quality than those commonly found in poorly exposed outcrops or in mine workings.

Trace minerals—such as the vivianite, pyrite, and siderite noted by Thayer (1970)—are usually used to infer paludal conditions in the absence of preserved coal. Where coal is present, however, the interpretive significance of such minerals is much reduced.

Palynology has proved exceptionally useful in dating Triassic coals and the associated deposits, but it is more limited in its value for inferring depositional environments. One major constraint on palynological studies is the possibility that indigenous pollen has been mixed with pollen blown in from nearby environments.

Currently, the most productive types of sedimentological studies of coal deposits have been those dealing with the sedimentology of the interbedded clastic lithologies. Definition of the morphologies of associated sandstones often allows the geologist to infer the paleofluvial regimes and thus the types of adjacent paludal environments. This is the type of approach taken by Ressetar and Taylor (1987). In the case of the Triassic coals, many of the reported associated lithologies are fine-grained lacustrine deposits. This supports the lake-margin/paludal (limnic) environments inferred by Robbins, Wilkes, and Textoris (1987). Other coals are found in association with fine-grained fluvial rocks and probably formed on poorly drained, low-gradient floodplains.

The position of many of the coal deposits near the top of tectonically influenced depositional cycles indicates that coals did not form at the distal margins of the alluvial fans. Nevertheless, the stratigraphic position of several of the coaly units above fluvial sediments and below lacustrine mudstones (figure 7-3) suggests that some coals formed in lake-margin coal-swamp environments that were facies equivalents of alluvial fans in one direction, and of lacustrine environments in the other. In this generalized model, a decrease in alluvial-fan sedimentation and an expansion of the fluvial and lacustrine environments could have produced the observed vertical sequences of lithologies.

8 / Lacustrine Deposits

Tremendous variability exists within both modern lacustrine environments and ancient lake deposits. Moreover, a time-dependent element of cyclic deposition is recorded in ancient lacustrine rocks but is often not apparent in modern environments. For these reasons, the criteria that have been used as diagnostic of lacustrine paleoenvironments in the Triassic–Jurassic rift basins often do not conform to models of lacustrine characteristics such as those listed by Collinson (1978) or Picard and High (1972). Table 8-1 lists some criteria that have been proposed as indicative of generalized lacustrine deposits and compares them with criteria that have been used as a basis for general lacustrine interpretations in the Triassic–Jurassic strata.

The characteristics of lacustrine deposits that have been emphasized by different authors as being diagnostic—and even the interpretations of what is generally agreed to be a diagnostic characteristic—often differ, depending on whether an author is trying to distinguish the lacustrine deposits from fluvial or from marine sediments. An example of this is the significance of lateral continuity of beds: Tucker (1978) suggested that rapid lateral and vertical facies changes in the Triassic of South Wales are definitive evidence for lacustrine sedimentation, as opposed to marine bedding that would be of greater lateral extent; in contrast, Clemmensen (1978b) argued that the lateral continuity of single beds for distances of up to 75 kilometers in the Greenland Triassic formations distinguishes them as lacustrine and distinct from the more variable floodplain deposits. Neither author is wrong: the difference in their assessments of the deposits in question only highlights the importance of interpreting sediments in context and underscores the variability in lacustrine environments and their deposits.

The term *lacustrine* covers such varied environments as flood-plain pud-

Table 8-1. **Examples of characteristics that have been used to infer lacustrine environments in rift-basin strata, compared to characteristics that have been listed for lacustrine facies models.**

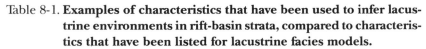

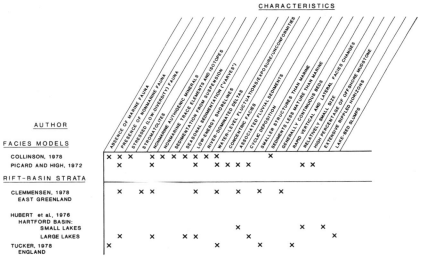

dles, structural features containing water over 300 meters deep, and shoreline embayments that have been cut off from the ocean. The variations of lake deposits form a continuum with both playa and marine sediments. Lakes may be huge or small, saline or fresh, deep or shallow; they may have waters that are mixed or stratified, anoxic or oxygenated; and they may deposit primarily clastic, organic, or chemical sediments. Almost all combinations of these end members and the intermediate conditions were present in the ancient rift-basin lakes.

The extent and significance of the Triassic–Jurassic lakes were not fully appreciated until the early 1960s (Van Houten 1962, 1964, 1965*a*, 1965*b*; Sanders 1968), but they have since been the subject of a disproportionate amount of the study of rift-basin strata. In spite of this, the exact causes of lacustrine conditions in the rift basins are still unproven. Proposed mechanisms include the following: increased precipitation during climatic trends toward humidity, such that lakes filled existing closed basins; primary tectonic damming by tilting of the basin floors and the consequent disruption of established through-drainage patterns; and secondary tectonic damming, such as the establishment of dams across through-drainage by the growth of large alluvial fans or the extension of lava flows.

Sanders (1968) suggested that, for the most obviously lacustrine strata in the Hartford basin, the extent of the deposits indicates lakes of a size that were too large to have been caused by a secondary damming of drainage, and most secondary discussion today revolves around primary tectonic or climatic causes. Olsen (1980) has suggested a compromise according to which the North American rift-valley lakes were created by unspecified tectonic events, while subsequent fluctuations in the lake levels responded to variations in rates of precipitation.

Several studies have noted that the thickest lacustrine strata are in fact near the faulted basin margins (Hartford basin: LeTourneau 1985a, 1985c. Newark basin: Turner-Peterson 1980; Olsen 1980. Culpeper basin: Lindholm 1978a, 1978b, 1979; Hentz 1985), suggesting that tectonic tilting of the basin floor and drainage reversals may therefore have been responsible for damming drainage and forming the lakes. Since many of the lacustrine deposits are associated with basalt flows, those lakes may have been a response to major tectonic reorganization that caused both volcanic activity and changes in basin-floor paleoslopes. Manspeizer (1981, p. 4-17) has suggested that, in the Newark basin, "subsidence along the border fault was about two to three times greater during the volcanic-lacustrine phase than during the earlier nonvolcanic redbed phase," which supports the idea that the lakes formed in response to changes in the tectonic regime. A few of the lakes, as recorded by the Lockatong Formation in the Newark basin and by the basal strata of the Kerrouchen basin, formed during periods of relative tectonic stability—possibly during initial stages of basin development, when faulting was minimal and less-localized subsidence was taking place.

The deposits discussed in this section are those of perennial lakes, and include shoreline, delta, and offshore deposits. The associated playa-lake deposits are covered in chapter 9. Although this discussion concentrates on the deposits of large rift-valley lakes, several authors (Sanders 1968; Hubert, Reed, and Carey 1976) have interpreted certain other deposits as being those of small floodplain lakes. The evidence for such interpretations includes fine-grained sediments that occur as thin, laterally extensive (kilometer-scale) beds without associated channels, and that contain such sedimentary structures as ripples (both oscillation and current), planar laminations, and starved ripples. These deposits are usually red, indicating oxygenated waters, and are often associated with evidence of intermittent subaerial exposure, such as desiccation cracks.

Gore (1985a) has suggested that a floodplain lake is recorded in the Deep River basin by a 2-centimeter-thick bed of "ostracodal limestone with disarticulated fish parts," with an overlying 2-meter-thick bed of fossiliferous red claystone. Such smaller floodplain lacustrine sediments have not been widely recognized, however, and they are not discussed further here.

Shoreline Deposits

Shorelines can be divided into a number of subenvironments. Those recognized in the rift-basin deposits include deltas, beaches, and mud and/or sand flats, as well as several overprinting processes that produced secondary erosional or reworking features. Deltaic deposits formed at the intersection of fluvial systems with lake margins. Gilbert deltas were created where the rivers carried coarse bed-load material, and lobate deltas formed elsewhere (due to the relatively low-energy wave systems in lakes). Beaches formed between deltaic centers during episodes of stable lake level. Both sandy beaches and pebbly beaches of reworked conglomerate have been described. In areas of lower relief, shoreline mud flats and sand flats formed.

Deltas

Possible Gilbert deltas have been described from the Lockatong–Stockton sequence in the Newark basin by Turner-Peterson and Smoot (1985) and by Smoot et al. (1985). These deposits are somewhat finer-grained than "typical" conglomeratic Gilbert deltas, and therefore have a lower foreset angle. The foreset dip angle decreases downdip, where the sandy foresets become interbedded with bottom-set mudstones. The coarsening-upward deltas are 3 to 4 meters thick overall, and the foreset laminae are composed of climbing ripples (Turner-Peterson and Smoot 1985). Fluvial topsets are seemingly absent, which has troubled the authors: it would be difficult to have deltas prograde into a lake without some attendant process that transported sediment out to the expanding delta front. These inferred delta sequences are often stacked, suggesting lake-level fluctuations; and Turner-Peterson and Smoot suggest that the fluvial topsets may have been eroded during lake-level low stands, although no other evidence has yet been presented that would bear on this problem.

River-dominated deltaic deposits have been suggested for two of the Moroccan basins. Brown (1980) inferred that lenticular sandstones (up to 150 meters thick) found at the top of a generally coarsening-upward sequence with interbedded organic-rich fine-grained sediments were distributaries and distributary-mouth bars interbedded with delta-plain and interdistributary-bay deposits. Lorenz (1976) suggested that smaller-scale lenticular sandstones (1 to 3 meters thick) interbedded with thin-bedded extensive mudstones were river-dominated deltas that prograded into a lake. In both basins, the sandstones are indistinguishable from fluvial-channel sandstones. They have basal scours and mudstone rip-up clasts, fining-upward grain-size profiles, and sedimentary structures that indicate decreasing flow velocities upward. The deltaic interpretation of these channels is based primarily on the lacustrine nature of the associated deposits, and at least in the second case, the sandstones could as easily be interpreted as truly fluvial lenses that had been superimposed on the lacustrine strata during low-water stages of the lake.

Beaches

Lacustrine beaches are thin and often are not preserved because of frequent fluctuations of lake level. Their sedimentary structures are similar to those of marine beaches, but they are significantly smaller. Beach deposits in the rift basins are usually recognized on the basis of good to very good sorting and extensive low-angle to planar-horizontal bedding (Tucker 1978; LeTourneau and Smoot 1985); local placer concentrations of heavy minerals have provided additional evidence for such interpretations (Turner-Peterson 1980). Lacustrine beach deposits cannot readily be distinguished from near-shore bars, which contain similar structures and occur within similar lithologic assemblages.

Brown (1980) suggested that sandy deposits containing concentrations of abraided plant and animal remains in Morocco may represent beach lags, and

Figure 8-1. Carbonate intraclasts in fine-grained sandstone (pencil for scale), interpreted as tempestites in shallow lakes (LeTourneau and McDonald 1985), Hartford basin, Connecticut. These thin beds are often associated with plant debris and molds of bivalve shells. (Photo by the author)

Clemmensen (1978*b*) interpreted flat-pebble conglomerates (mudstone rip-up intraclasts from associated carbonate mud flats) in Greenland as storm-lag deposits on beaches. LeTourneau and McDonald (1985) have interpreted similar intraformational conglomerates in Connecticut as storm-generated "tempestites." These thin beds (figure 8-1) contain both well-rounded and angular intraclasts, as well as large (fragile) plant fragments and small disarticulated fish parts; they are associated with small-scale hummocky crossbeds. The features of the thin beds indicate high and mixed energy levels, while hummocky crossbeds suggest storm-surge waves.

Beach gravels, which formed during lacustrine high stands, have also been recognized within alluvial-fan deposits (Tucker 1977; LeTourneau 1985*b;* LeTourneau and Smoot 1985; Smoot et al. 1985). At such times, the fans prograded directly into lakes, and shorelines were located on the fan slope, where wave action reworked the existing alluvial-fan deposits. These reworked, secondary deposits are distinguished from those resulting from fan and fluvial depositional processes by the clast-supported nature of the conglomerates (figure 8-2), which show a well-sorted, infiltrated matrix, and by the associated sandy deposits with parallel bedding and oscillation ripples.

Figure 8-2. Layer of well-sorted, clast-supported conglomerate, interpreted as shoreline (beach) deposits, interbedded with poorly sorted, matrix-supported conglomerates, South Wales. The well-sorted beds are laterally discontinuous on the scale of meters, suggesting the narrow scale and ephemeral presence of lacustrine shorelines. (Photo courtesy of M. E. Tucker)

Tucker (1978) has documented instances where high-stand Triassic shore-lines abutted limestone bedrock slopes and produced several wave-cut terraces in Wales. The resulting scarp shed angular clasts onto a scree slope, and the clasts were then reworked into beach gravels.

Another possible beach or near-shore deposit, in the Kerrouchen basin in Morocco, consists of well-sorted millimeter-scale basalt sand with a sparry carbonate cement (figure 8-3). This meter-thick black-sand bed, eroded from contemporaneous local basalt flows, occurs within an extensive mud-stone facies that may be attributable to playa- or marine-influenced environments, as discussed in chapters 9 and 10.

Mud Flats

Shoreline mud flats have been recognized in several of the rift basins. These were apparently the result of local low-relief shorelines, or of occasional lacustrine low stands; they are composed of either carbonate or siliciclastic

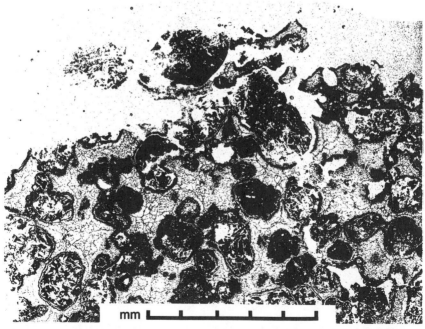

mm ┗━━┻━━┻━━┻━━┻━━┻━━┛

Figure 8-3. Possible beach deposit, Kerrouchen basin, Morocco. Well-sorted sand grains of rounded basalt fragments in a sparry carbonate cement. (Photo by the author)

muds. Although they are commonly massive, both types contain remnant oscillation ripples, track and trail trace fossils, and desiccation cracks. The siliciclastic mud flats of Greenland are interpreted to be initial flood deposits that were subsequently reworked in the lacustrine environment, while the carbonate muds are inferred to have been derived primarily from shallow-water algal blooms (Clemmensen, 1978*b*).

Tucker (1977) has interpreted evaporitic mudstones and carbonates that often contain stromatolite structures as sabkha-type shoreline mud flats (figure 8-4A) in South Wales. These mud flats are transitional between coarse basin-margin clastics and nonevaporitic mudstones that are interpreted as offshore lacustrine in origin. Teepee structures (figure 8-4B) and caliches also formed on the mud flats. (As noted in chapter 10, these English deposits may not be entirely nonmarine.)

LeTourneau and McDonald (1985) have also suggested that the western shorelines of lakes in the Hartford basin included widespread evaporitic mud flats, although here the concentration of evaporite minerals was diluted by influxes of fine-grained clastic material. In this basin, the mud flats that developed on the western, low-relief shorelines can be contrasted with the conglomeratic shorelines that formed in places where the lake encroached onto alluvial fans to the east across the basin.

Shallow-water Deposits

The Triassic-Jurassic rift-valley lakes had considerable expanses of shallow-water environments in which muds, silts, sands, and sometimes carbonates were deposited. In some places, shallow-water deposits constitute the major portion and the maximum development of the lacustrine sequences. In others, these sediments formed a transitional facies that existed as a shallow-water swath between shoreline and offshore environments. The geochemistry of the lake waters and the rates of clastic influx determined whether clastic or chemical sedimentation predominated in these areas, as well as what form of carbonate deposition took place.

Carbonates

Shallow-water lacustrine carbonates are present in different rift basins in different forms. The most commonly mentioned (though poorly described) form is micritic limestone, usually of unspecified origin but often with suggestions of algal laminations or freshwater charophyte oogonia (algal reproductive structures). Some contain a few recognizable freshwater invertebrate fossils (Newark basin: Turner-Peterson 1980). These limestone beds may range in thickness from 10 centimeters to 4 meters and locally may be seen to grade laterally into more definitive types of carbonate deposits.

Some of these limestones contain disconformities, desiccation cracks, and vertebrate tracks, suggesting frequent subaerial exposure as carbonate mud flats (Culpeper basin: Gore 1985*b;* Newark basin: Manspeizer 1980). Tucker (1975, 1978: South Wales) and Clemmensen (1978*b:* Greenland) have described fenestral fabrics, caliches, and teepee structures (figure 8-4B) in some of the limestones—signs of subaerially exposed, evaporative carbonate mud flats. Other limestones, or often the same beds, contain oscillation ripple marks and either subaqueous burrows or tracks and trails indicating relatively shallow waters (Newark basin: Manspeizer 1980; Greenland: Clemmensen 1978*b*).

Stromatolitic limestones, which require clear, sediment-free waters, are found in a few of the rift basins. Clemmensen (1978*b*) has described small stromatolites from Greenland as 5- to 15-centimeter-thick laterally linked hemispherical heads with common overhangs, and locally as having elongate heads that are interpreted to have formed parallel to "wave scour." One of these 5-centimeter-thick stromatolitic limestones can be traced laterally for 75 kilometers, indicating the minimum extent of uniform shallow-water conditions along the axis of the basin, as well as the widespread absence of clastic input to the basin at that time. This is an exceptional occurrence within the rift basins.

Slightly more common, although still infrequent, is the presence of oolitic limestones. The best examples of ooids have been reported from the Culpeper basin (Carozzi 1964; Young and Edmundson 1954). Carozzi compared these to the ooids forming in the modern Great Salt Lake and inferred a shallow-water origin. Ooids generally require agitated shallow water and associated algal growth in order to form.

The Culpeper basin oolite is 15 to 20 centimeters thick, and there are indications that partially cemented beds of ooids were sometimes ripped up

Figure 8-4 *(facing page)*. **Characteristics of evaporitic mudflats of South Wales. (A) Nodular dolomite that replaced anhydrite—a rock fabric and mineralogy that is comparable to modern Trucial-Coast sabkha deposits (on the Persian Gulf)—and has been used to infer large, inland, nonmarine sabkhas associated with a large lake (Tucker 1978). (B) Small teepee structure (total height about 20 centimeters) in mud-flat carbonates. Associated with stromatolites, desiccation cracks, fenestral fabrics, ostracods, and laminated to cryptalgal carbonates, these rocks are interpreted as "shallow subaqueous deposition, interrupted by periodic exposure, of a lacustrine shore-zone carbonate mudflat. . . ." (Tucker 1978). (Both photos courtesy of M. E. Tucker)**

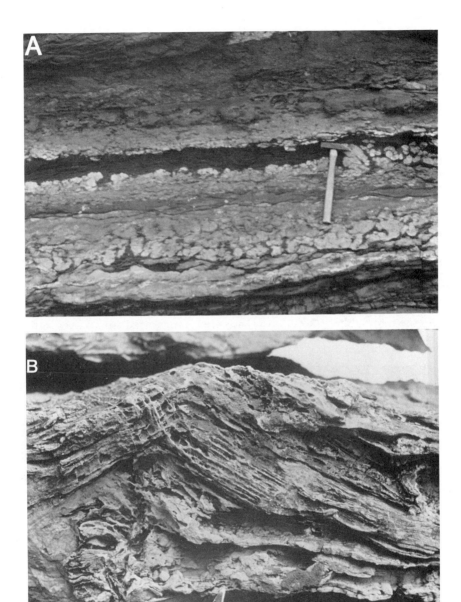

to form intraclasts during high-energy storm events. At other times, subaerial erosion formed unconformities within cemented oolitic beds, truncating grains and creating desiccation cracks.

Oncolitic carbonates also were formed in rift-basin lacustrine environments—sometimes in siliciclastic depositional systems (figure 8-5), and occasionally in dominantly carbonate systems (Tucker 1975). Formed initially as algal coatings on grains, the coatings grew, entrapping other sediment. Wave or other agitation was apparently less regular in this subenvironment than in the one responsible for forming ooids, as oncolitic clasts are much more irregular in shape.

Clastics

Shallow-water clastic sediments occur as lake-bottom and depositional-slope deposits in the rift basins. The latter commonly formed where rates of clastic sedimentation were high (usually near deltas) and are differentiated from lake-bottom sediments by the common occurrence of contorted bedding caused by slumping on relatively high-angle, oversteepened slopes (figure 8-6). Hubert, Reed, and Carey (1976) measured slump axes in the Hartford basin and inferred lake-bottom paleoslopes, but—as pointed out

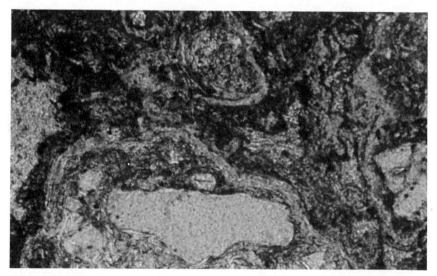

Figure 8-5. Roughly concentric oncolite structures binding sand grains (probably deposited in a shallow lacustrine environment), Kerrouchen basin, Morocco. Long axis of the largest sand grain is approximately 1 millimeter. (Photo by the author)

by Hentz (1985)—oversteepened slopes are usually local lake-margin depositional phenomena and are not related to the regional, tectonic paleoslopes in these basins.

Shallow-water lake-bottom deposits are common and are usually recognized by the presence of oscillation and interference ripple marks (figure 8-7), as well as by good sorting. Deposited in shallow, oxygenated waters, they are commonly red, but gray sediments also occur in association with carbonaceous debris (LeTourneau and Smoot 1985). Algal mats and bivalves have been found in some of these shallow-water deposits (LeTourneau and McDonald 1985). The strata containing these fossils range from claystones to coarse sandstones. They are often burrowed; the trace fossil *Scoyenia* (described in chapter 6) has been ascribed to this environment (Turner-Peterson 1980). Burrowing disrupts primary sedimentary structures, and these deposits may be entirely structureless due to bioturbation.

Where sedimentary structures remain, they include small-scale wave-generated hummocky bedding with quiet-water mud drapes (Hentz 1985), extensive graded beds (Tucker 1978), and parallel-horizontal laminations formed by settling from suspension. These last probably mark the transition into deeper waters, below wave base.

Shallow-water clastic beds may be as extensive as some of the carbonates. Clemmensen (1978b) has suggested that one 4-meter-thick bed of rippled sandstone can be traced for 70 kilometers along the axis of the same basin in Greenland in which the extensive stromatolitic limestone was found.

Figure 8-6. Slump fold in a turbidite sandstone bed from lake-slope deposits, Culpeper basin, Virginia. (Photo courtesy of T. F. Hentz)

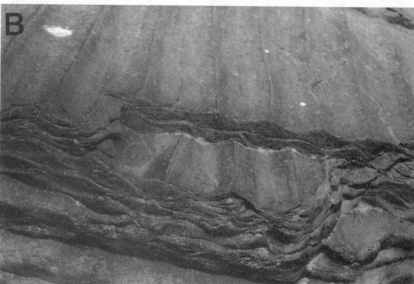

Figure 8-7. Ripple marks in shallow lacustrine deposits, Hartford basin, Massachusetts. (A) Extensive rippled bedding plane. (B) Close-up of ripples, showing the internal cross-laminations of current ripples and the external morphology of oscillation ripples. The width of the area shown in the photo is about 15 centimeters. (Both photos by the author)

These shallow-water deposits were sometimes subject to secondary processes during episodes when lake levels were lowered and subaerial exposure ensued. The most common process was the formation of desiccation cracks; but fluvial channels, too, were locally superimposed on the lacustrine deposits (Hentz 1985). Elsewhere repeated episodes of wetting and drying led to distinctive teepee structures and crumb fabrics of millimeter-scale brecciation in the mudstones (Smoot and Olsen 1985).

Deep-water or Offshore Deposits

Many of the shallow-water deposits have offshore facies equivalents. These are usually inferred to have been deep-water depositional environments on the basis of the presence of turbidites, the absence of wave/oscillation ripple marks, the great lateral extent of many such deposits, and the presence of unoxidized organic material. The last type of evidence mentioned above is usually used to infer anoxic conditions in stratified lakes and to imply "deep" lakes.

The presence of shoreline deposits on the slopes of alluvial fans indicates that depths of at least some of the lakes were as great as the vertical relief between the toe of the fan and the location of the shoreline; unfortunately, this can rarely be reconstructed in these strata.

Water depths are difficult to reconstruct. Hubert, Reed, and Carey (1976: Hartford basin) calculated that a hypothetical ¼ degree of basin-floor/lake-bottom slope (away from the alluvial fans) across a 20-kilometer-wide basin would produce a maximum water depth of about 80 meters, but they assumed a westerly basin-floor paleoslope. However, LeTourneau (1985a, 1985c) has recently suggested that the basin-floor paleoslope probably dipped to the east (toward the fans) during lacustrine sedimentation, and therefore that the deepest parts of the lakes were probably at the toes of the fans (figure 8-8).

Olsen (1985a) offered, as an indication of the size of some of the lakes, observations of the continuity of some deep-water deposits over thousands of square kilometers in the Newark basin, with minimal facies changes (figure 8-9). He felt that such continuity implied a probable water depth of over 100 meters.

Although the black ("anoxic") mudstone facies that many authors have cited as the primary evidence for deep lakes may occur in stratified waters as shallow as 4 meters (for example, in modern Lake George: Robbins 1985b), other qualitative evidence, as noted above, suggests that many of the black laminated mudstones are indeed relatively deep-water facies.

Black, organic-rich, often calcareous mudstones are the most commonly recognized deep-water lithology. These are finely laminated, with no evi-

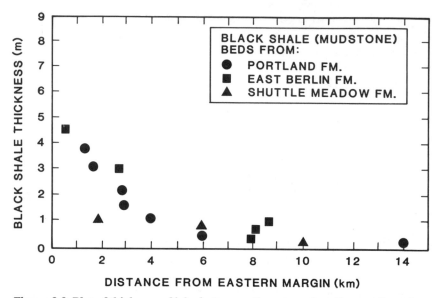

Figure 8-8. Plot of thickness of lake-bottom sediments against distance from the main basin-margin fault, showing that the lacustrine strata thicken toward the fault and that, therefore, the regional paleoslope at the time of deposition of each of these three deposits was probably toward the fault. (From LeTourneau and McDonald 1985, Figure 7. Reprinted with the permission of the Connecticut State Geological and Natural History Survey)

dence of currents or burrowing (figure 8-10), and locally contain well-preserved, articulated fish fossils. The horizontal, millimeter-scale laminations suggest deposition below wave base, possibly during annual seasonal changes in water chemistry due to overturn of stratified lakes (Hentz 1985). Hubert, Reed, and Carey (1976) noted alternate laminations of black mudstone and dolomite, suggesting that the mudstone layers were deposited during rainy seasons by the influx of clastic detritus and concurrent high rates of organic production, whereas the dolomite laminations were precipitated during drier seasons. These authors also argued that the locally abundant fish fossils in this lithology represent large-scale fish kills during periodic overturn of the stratified lake waters. Inhospitable bottom waters are often indicated by the absence of burrowing, the absence of scavengers that would have destroyed the fish before fossilization, the unoxidized organic content of the black mudstones, and the common presence of authigenic pyrite. These waters may have been anoxic and/or hypersaline as a result of chemical or thermal stratification.

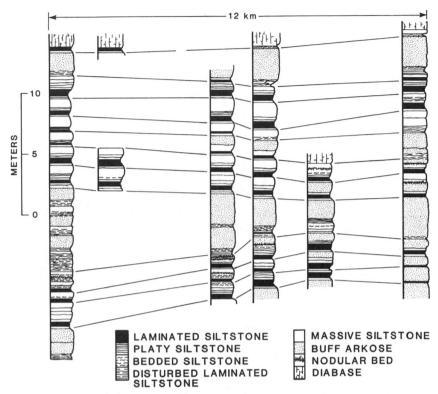

Figure 8-9. Inferred lateral continuity of lacustrine strata in the Newark basin, New Jersey. (From Olsen 1980, Figure 33. Reproduced with the permission of the New York Geological Association)

Porter and Robbins (1981) and Gore (1986) have recently suggested that other black mudstones originated as organic-rich pelleted mudstones that were deposited in bioturbated and oxygenated environments—the diagnostic criteria having been largely destroyed during compaction. Porter and Robbins (1981) found evidence that the organic matter in the black mudstone facies of the Deep River basin had been produced as the fecal pellets of zooplankton (figure 8-11) and that much of the silty fraction of this facies is organically derived. Gore (1986) has reported remnant features in black mudstones from the Culpeper basin that are composed of abundant ostracod valves and fecal pellets. These data indicate that an environmental model of deep, stratified lakes may be inappropriate for some of the black mudstone facies.

Figure 8-10. Deep-water (basin-plain) lacustrine lithofacies. (A) Laminated black mudstone, Hartford basin, Massachusetts. Note the fine, discontinuous laminations, indicative of very low-energy conditions. (B) Extensive, unrippled planar bedding planes of same facies, Deerfield basin, Massachusetts. (Both photos by the author)

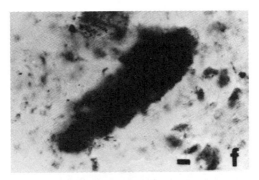

Figure 8-11. Fossil copepod (aquatic microcrustacean) fecal pellets from the lacustrine Cumnock Shale. Bar scale is 10 micrometers. (Photo courtesy of E. I. Robbins, from Porter and Robbins 1981, Figure 1. Reprinted with the permission of *Science*, copyright 1981, by the American Association for the Advancement of Science)

Thin, graded, sandy beds of probable turbidite origin (figure 8-12) are locally associated with the black mudstones, however, suggesting that these mudstones at least were deposited in relatively deep waters. The turbidites may be of siliciclastic composition (Sanders 1968) or may be made up of resedimented shallow-water carbonates (Lee 1977; Gore 1985b). Hentz (1985, p. 102-4) has observed graded, conglomeratic "gravity flows" interbedded within "basinal mudstone and turbidite" deposits of the Culpeper basin.

It is usually difficult to determine the lateral extent and variability of the deep-water lacustrine deposits. Olsen (1980) has suggested lateral continuity over a few tens of kilometers (see figure 8-9) and Demicco and Gierlowski-Kordesch (1986, Figure 8) have presented similar data for the Hartford basin. Although thicknesses vary significantly, sedimentary packages can be correlated well across the scale of a few kilometers. This indicates a minimum extent for these lacustrine environments and may give an indication of differential subsidence within the basins.

Other types of offshore lacustrine deposits have been reported. Clemmensen (1978b) suggested that laterally extensive beds of laminated green mudstones, siltstones, and sandstones in the Central East Greenland basin represent deposition from suspension in an "open lake" environment. Some of these deposits are dark gray, and (in the absence of burrowing) the bottom-water environment was inferred to have been anoxic. In South Wales, Tucker (1977) has interpreted parts of the red dolomitic mudstone of the English Triassic to be offshore lacustrine sediments. These deposits apparently underwent periodic subaerial exposure during a lacustrine regression, when nonmarine sabkha conditions are inferred to have produced horizons of

Figure 8-12. Rhythmically laminated sequence (in the Culpeper basin, Virginia) of mudstone and calcilutite containing a thicker sandstone turbidite deposit, from a lacustrine basin-plain depositional environment. (Photo courtesy of T. F. Hentz)

gypsum nodules. Much of this formation has been interpreted as a marine-influenced deposit by other authors; it is discussed further in chapter 10.

Geochemistry of Lake Waters

The geochemistry of the rift-valley lakes is partially recorded by the fossils and authigenic minerals commonly found in such deposits. In general, lake waters that were both saline and alkaline are suggested, but significant variability existed. Lake geochemistry is dependent on source-rock mineralogy for geochemical components, and on climate and basin drainage patterns for the mechanisms and degree of concentration of the supplied components.

In some of the lakes (Greenland: Clemmensen 1978b) extensive stromatolites and a rich trace-fossil fauna indicate nonstressed, probably oxygenated, nonsaline waters. Clemmensen noted that, significantly, stromatolites and bioturbation begin to appear at the same stratigraphic horizon, suggesting a trend toward decreasing salinity. However, halite pseudomorphs and trace fossils occur together in some of the facies, indicating temporally or locally variable conditions of salinity.

Invertebrate fossils that suggest fresh and brackish water are common in some of the lacustrine deposits. Most commonly reported are conchostracans (bivalved crustaceans), which are of indefinite significance. Van Houten (1962) has reported fresh-water bivalve fossils in the Newark basin, noting that they are restricted to margins of the basin and thus (presumably) to shallower, well-oxygenated waters. McDonald (1985) has suggested that the apparent absence of invertebrate fossils in many of the basins is more often a consequence of lack of preservation than a function of inhospitable waters. He has reported an assemblage of invertebrates from probable near-shore environments of the Hartford group.

At the other extreme of the lake-water geochemistry spectrum are the analcime-rich *argillites* (indurated, nonfissile mudstones) of the Lockatong Formation in the Newark basin, described by Van Houten (1962, 1964, 1965a, 1965b). Analcime, an authigenic, sodium-rich zeolite, forms a significant component of this formation and was precipitated from highly alkaline, sodium-rich lake waters. It is commonly associated with pseudomorphs of glauberite, a sodium-calcium sulfate evaporite that grew downward from the mud-brine interface as rosette crystals within the unconsolidated analcime muds. The lake-water sodium was evidently derived from a source area that contained common sodium feldspars, and it was concentrated in a basin characterized by internal drainage and high evaporation. Repetitive sequences of increasing concentration of ions in the lake waters are recorded in the repeated vertical sequences of minerals with increasingly high percentages of sodium and carbonates. Similar argillites are present in parts of the Hartford basin, although they occur in significantly less volume.

In other basins, intermediate levels of alkalinity prevailed. In Massachusetts, Stevens and Hubert argued that "ferroan dolomite in gray mudstone, together with albite and ferroan dolomite cement in sandstone, suggest a closed-exit alkaline lake ... " (1980, p. 110). Tucker (1978) suggested that, in South Wales, the assemblage of clay minerals that includes illite, corrensite, swelling chlorite, and montmorillonite is typical of deposits of alkaline, hypersaline, magnesium-rich lake waters in an arid climate. A magnesium-rich environment was also inferred by April (1981) for some of the lacustrine deposits in the Hartford basin. The nonlacustrine clay assemblages in this basin consist predominantly of detrital illite, whereas the lacustrine clay assemblages contain significant amounts of authigenic and diagenetic types of chlorite and smectite that could have formed in the presence of magnesium-rich lake waters and/or pore fluids.

Evidence of evaporites is locally common in the lacustrine deposits, indicating high levels of salinity in some lake waters. Such evidence ranges from scattered pseudomorphs of halite hopper crystals in the black shale facies, indicating saline stratified waters (Parnell 1983; Thayer 1970), to beds of gypsum, anhydrite, and halite that are transitional between lacustrine and playa deposits (Tucker 1978; LeTourneau and McDonald 1985).

Facies: Patterns and Controls

Sequences or cycles of lacustrine facies patterns in the Triassic–Jurassic rift basins were first noted by Van Houten (1962, 1964) in the Lockatong Formation in the Newark basin. Since then, cyclic patterns of lacustrine sedimentation have been reported from most of the other North American basins, and although some of them do not possess the detailed patterns originally described by Van Houten, Olsen (1985b) has recently proposed that such sequences be generically called *Van Houten cycles*. As noted in Part I, these patterns are usually assumed to record climatic fluctuations.

Type Cycles

As originally described (1962), Van Houten's well-documented cyclic deposits are of two types. Cycles of primarily chemical deposits (authigenic analcime and dolomite) constitute two-thirds of the upper Lockatong Formation, increasing in thickness and in relative abundance toward the center of the basin. Cycles of detrital composition make up the remainder of the formation. The detrital cycles are about 5 to 7 meters thick and consist of a lower unit of laminated pyritic black mudstone, grading up into calcareous silty mudstone with contorted bedding (figure 8-13). The chemical cycles are only 2 to 3½ meters thick. They, too, begin with a lower black mudstone facies, but one that contains significantly more calcareous material than the detrital cycles' mudstone does, grading up into disrupted to massive analcime- and dolomite-rich argillite.

These cycles are asymmetric, in that they record unidirectional progressions that start and end abruptly at opposite ends of the depositional spectrum. The lower black mudstone facies in each case is interpreted as a stagnant-water, anoxic environment associated with a rapidly filled, stratified, but not necessarily deep lake, whereas the upper, disrupted mudstones record ephemeral, playa-lake conditions (Van Houten 1964).

Olsen (1980, 1985a) has since suggested that there is commonly a thin initial transgressive sequence below the black mudstone facies. It may be composed of platy to massive mudstones deposited on mud flats or in fluvial environments, or it may only be a less-desiccated final stage of the top deposits of the last sequence. This additional facies gives the sequences a slightly more symmetrical character.

The disrupted to laminated mudstones above the lower black mudstones

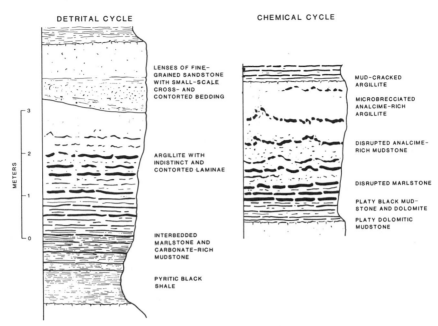

Figure 8-13. Cyclic lacustrine deposits, as originally recognized by Van Houten. These are schematic representations of the numerous measured cycles in the Lockatong Formation, Newark basin, New Jersey. (After Van Houten 1962, Figure 3. Reproduced with the permission of the *American Journal of Science*)

are interpreted as suspension (detrital-cycle) or precipitation (chemical-cycle) deposits in an increasingly shallow lake. These sedimentary rocks are inferred to record a drying-up and emergent process, rather than a regressive one, because the vertical mineralogical sequences are similar across the basin, varying only in thickness. Detrital deposits were more common on lake margins and formed primarily during climatic intervals of increased precipitation of under basin conditions of through-drainage, whereas chemical sediments were predominant in the basin center, where brines concentrated during relatively arid episodes or under closed-basin conditions.

According to Van Houten (1964), the disruption of the laminations in the upper parts of the cycles may be due to burrowing in the detrital deposits, to growth of evaporite crystals in the soft muds of the chemical sediments, or to postdepositional compaction and dewatering. More recently, Smoot and Olsen (1985) have suggested that most of this fabric—including desiccation-cracking and microbrecciation—is due to soil-formation processes that took place as the lake contracted and the sediments were exposed to repeated wetting and drying in the final playa setting.

Other Cycles

After these lacustrine cycles had been described, defined, and interpreted by Van Houten, other lacustrine cycles began to be recognized. In the Hartford basin, Hubert, Read, and Carey (1976) and Hubert et al. (1978) described several lacustrine cycles within the East Berlin Formation (see figure 4-6); LeTourneau (1985*a*, 1985*c*) has recognized similar sequences in the lower Portland Arkose. These deposits were associated with basalt flows and a less-stable tectonic environment than were those of the Lockatong Formation. Because of the fair degree of vertical symmetry, however, and possibly because of Van Houten's previous climatic interpretations, they have been interpreted as resulting from the gradual filling and emptying of rift-valley lakes during changes from more arid to more humid and back to more arid climatic regimes. As in the case of the Lockatong deposits, the gray mudstones were interpreted as shallow-water deposits, and the black mudstones were explained as deeper-water/offshore deposits in stratified conditions. Thus, a relatively symmetrical transgressive–regressive cycle was inferred. The associated sandstones occur at the extremes of the sequence and were taken to be the records of shallow-water/shoreline environments.

Recent work by Demicco and Gierlowski-Kordesch (1986) has suggested that at least fifteen lacustrine cycles occur within the upper 100 meters of the East Berlin formation—more than were originally recognized—and that many of them more closely resemble the Lockatong cycles in their asymmetry (figure 8-14). Demicco and Gierlowski-Kordesch reconstructed a sequence composed of a lower sandstone/mudstone that records a rapid transgression, followed by a laminated black mudstone (interpreted as the deposits of the maximum lacustrine development), and capped by a regression/playa-lake suite of mudstones and sandstones. Demicco and Gierlowski-Kordesch apparently did not find shoreline deposits at the top of the cycles, as had Hubert, Reed, and Carey (1976). Instead, their upper deposits consist of disrupted mudstones formed in the playa environments discussed in chapter 9.

Olsen (1980, 1985*a*, 1985*b*, 1986) has suggested that asymmetrical cycles are common to many of the North American rift basins, but he has expanded the boundaries of the concept to include fluvial deposits. Some of Olsen's described cycles are up to 70 meters thick, the upper two-thirds of which are fluvial in origin, and the resultant "cycles" are not readily distinguishable from the local fluvial–lacustrine intertonguing of entirely different origin. Other authors, following Olsen's lead, have applied the cycle concept to 150-meter-thick sequences that grade from deep-lacustrine to fluvial basin-fill sequences, comparing them directly with the 2- to 7-meter-thick Lockatong cycles.

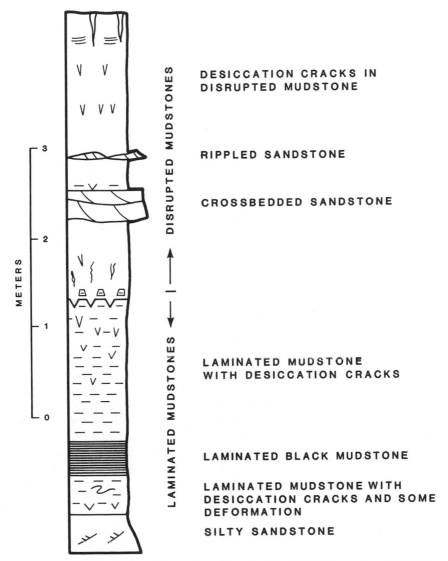

Figure 8-14. Cyclic lacustrine-to-playa deposits, East Berlin Formation, Hartford basin, Connecticut; contrast this type of cycle with the one originally reconstructed for this formation by Hubert, Reed, and Carey 1976 (see figure 4-6). (After Demicco and Gierlowski-Kordesch 1986, Figure 9A. Reproduced with the permission of *Sedimentology*)

This larger scale of "cycles" approximates the 100- to 125-meter scale of Van Houten's "long" Lockatong cycles, but the processes, products, and interpretations are entirely different. Van Houten recognized that the ratio of detrital to chemical cycles in the Lockatong Formation varied systematically vertically through the formation, and he suggested that such broad-scale variation was also a function of climatic fluctuations—but on a much broader, second-order scale.

Olsen (1986) has also inferred that there are at least three orders of cyclicity to the lacustrine sedimentary strata in the North American rift basins. Since the reconstructed periodicity of these cycles correlates roughly with the calculated periodicity of minor changes of the earth's rotation, Olsen suggests that the Triassic–Jurassic climate changed cyclically as a function of these events and that cyclic variation is recorded in the asymmetric lacustrine cycles. Still, no definitive proof has been found for the climatic theory, and although such cycles are "most easily explained by changes in precipitation and cloud cover" (Olsen 1986, p. 843), alternative hypotheses have not been explored.

Associations with Nonlacustrine Deposits

In addition to being interbedded with each other, deposits of the different lacustrine environments may be interbedded with any and all of the non-lacustrine deposits found in the rift basins. The Lockatong Formation, consisting primarily of stacked lacustrine cycles, is a notable exception.

Black "deep-water" mudstones may be interbedded with alluvial-fan conglomerates (LeTourneau 1985a), with paludal coals (Robbins, Wilkes, and Textoris 1987), and/or with evaporitic sabkha sequences (Tucker 1978). Some lacustrine facies vary rapidly laterally, while other thin lacustrine beds extend for tens of kilometers. Facies variations may differ significantly from one side of a basin to the other, depending on the local tectonics; in the Newark basin, Olsen (1980) has schematically diagrammed a black mudstone sequence that is thick and contains turbidites near the northwestern, faulted margin, but that thins and interfingers with shallow-water siltstones toward the opposite, unfaulted margin.

Lacustrine variability is too great to allow detailed discussion here. It is enough to note that, whenever a rift valley became subject to internal drainage, and/or whenever climatic precipitation exceeded water loss from the basin, lakes formed and lacustrine deposits became intimately interbedded with the sediments of whatever preexisting environments were present in the basin.

Evidence from Basalts

Basalt flows were limited to the Moroccan Triassic basins and to the North American basins from Virginia to Nova Scotia. Because basalt lavas are fluid and form deposits with relatively flat upper surfaces, they can sometimes be used to determine paleotopography. They also interact with water to form pillows, and thus can be used in a gross sense to infer the presence or absence of significant bodies of water at the time of extrusion. In addition, they commonly disrupt drainage as they flow across a valley floor, and the deposits that result from the disrupted conditions can sometimes indicate preflow environments. Although basalt flows are not sedimentary deposits, evidence from them may be useful in reconstructing depositional environments.

The lowest of the three basalt flows in the Hartford basin is commonly pillowed (see figure 3-4), although the identifiable underlying sediments are usually nonlacustrine. Hubert et al. (1978) have suggested that the lacustrine environment had not existed long enough prior to the basalt extrusion to have left any sedimentary record. The lower contact of the basalt with the sediments often consists of 6 to 7 meters of volcanic/sedimentary agglomerate, and a thin lacustrine sedimentary record could have been destroyed during the formation of this agglomerate. Father north, in the Deerfield basin, pillow basalts overlie 10 meters of lacustrine shales (Stevens and Hubert 1980).

To the south, Van Houten (1969) has noted pillows in the *second* flow unit within the lower multiple-flow basalt formation (First Watchung) in the Newark basin. Van Houten inferred that the first flow unit "deranged the regional drainage," and that the second flow "became pillowed where it encountered local lakes" (p. 328) ponded by the first flow.

In Morocco, most of the basalt flows occur high in the stratigraphic section. They are intercalated with evaporitic mudstones, and neither the mudstone facies nor the basalts are restricted to the confines of the rift basins. In this setting, the limited lateral extent of pillowed horizons is useful evidence in arguments over the origin (marine versus sabkha) of these deposits, as discussed in chapter 10.

Mattis (1974, 1977) has reported that these Moroccan basalt flows disrupted fluvial drainage and created local lakes within the rift basins of the High Atlas area, much as they are thought to have done in the Newark basin. The evidence includes local pillowing (lateral extents of only a few tens of meters) within flow units overlying the lowest flow unit. Thin beds of carbonates often directly overlie the flows. These carbonates contain stromatolitic beds that conform to the irregular upper surface of the flows (figure 8-15). They also contain ostracods and grade laterally and vertically into clastic fluvial redbeds.

Figure 8-15. Stromatolitic carbonates deposited in small floodplain lakes that formed when basalt flows disrupted local drainage, Central-N'Fis basin, Morocco. (A) Irregular and laminated carbonates conforming to the irregular upper surface of a basalt flow, overlain by red mudstones. (B) Close-up of algal laminations within the carbonates. Specimen is about 7 centimeters across. (Photos courtesy of A. F. Mattis)

Mattis has interpreted these as deposits of small lakes formed on the uneven surfaces of the basalt flows. Similar deposits occur in the Middle Atlas region, where up to four superimposed basalt-flow units occur, each with a thin overlying lacustrine carbonate bed (Lorenz 1976).

Summary

Triassic-Jurassic rift-valley lakes encompassed a variety of subenvironments and geochemical conditions. The inception of many of the lacustrine environments may have been tectonically controlled: most lacustrine deposits record relatively rapid transgressions and are associated with episodes of volcanism. Once the lakes had formed, lake levels, lake-water geochemistry, and the character of the resulting fine-scale lacustrine deposits are believed to have been responsive primarily to climate—both to long-range fluctuations and to shorter seasonal variations. Thicker regressive and/or drying-up sequences record slower disappearances of the lacustrine environments. Many of the lakes were stratified and contained anoxic bottom waters during their maximum development, but there are as yet no definitive measurements of water depths.

Lacustrine sediments are intimately interbedded with deposits from a wide variety of other environments, but they often formed their own distinctive vertical sequences of lithologies and sedimentary structures during transgressive/regressive fluctuations. Although they have been the focus of many recent sedimentological studies, the lacustrine sediments and their origins are still not completely understood.

Data Collection and Interpretation Techniques Illustrated by Studies of Rift-Basin Lacustrine Deposits

Lacustrine deposits exhibit a number of classic and generally well-understood primary sedimentary structures, such as graded turbidite sequences, oscillation (wave) ripple marks, and slump structures. The offshore deposits have good preservation potential, although the sedimentary structures within them may be destroyed by bioturbation or by secondary soil-forming processes during periods of subaerial exposure. Shallow-water and shoreline deposits are thin to begin with, however, and are subject to erosion during storms and low-water stages. Thus, an understanding of primary depositional structures, as well as of superimposed secondary postdepositional structures and erosional processes, is necessary for the reconstruction of lacustrine environments.

The orientation of directional sedimentary structures, such as current ripples and slump structures, is evidence for paleogeography. Hubert, Reed, and Carey (1976) have used these structures in the Hartford basin to indicate currents created by paleowinds and paleobasin slope, respectively. Such data should be used cautiously, however. Lakes are often closed systems, and therefore a current in one direction requires return flow in another; signs of such return flow have not yet been recognized in Connecticut. The slump structures may be more indicative of locally oversteepened subaqueous depositional slopes near deltas than of regional paleoslopes, as suggested by Hentz (1985) for the Culpeper basin.

Once the assemblages of sedimentary structures within individual beds have been interpreted, existing vertical sequences of repeated assemblages and lithologies may be recognizable. Because lacustrine environments are strongly influenced by climate, climatic fluctuations should be recorded in the sedimentary patterns.

The existence and composition of authigenic minerals constitute the primary evidence for reconstructing lake-water composition and (indirectly) climate. The uncommon minerals, glauberite and analcime, that abound in the Lockatong Formation are reliable indicators of an alkaline lake in an evaporative setting, as Van Houten (1962, 1964, 1965a, 1965b) has shown. The occurrence of molds of halite crystals in some of the black mudstone facies that also contain authigenic pyrite suggests saline lake-bottom waters.

Fossils and trace fossils are not as common in the Triassic–Jurassic rift-basin deposits as they are in other lacustrine settings, but they are being found with increasing frequency and are more common in the lacustrine facies than in deposits of the associated environments. The mode of life of related modern animals is the best existing clue to the conditions of fossil deposition, although Olsen (1980) has noted that the conchostracans whose forebears were common to many of the large Triassic–Jurassic lakes are today usually restricted to much smaller (and often ephemeral) lacustrine environments. The presence of bivalves and stromatolites, however, is indicative of oxygenated, probably sunlit, and relatively sediment-free waters. The occurrence of fish fossils in the black mudstone facies is good evidence of stratified lake waters: fish were able to live in the oxygenated upper waters, yet were well preserved after death in the anoxic bottom oozes.

The significance of trace fossils is usually more ambiguous, but Bromley and Asgaard (1979) have listed useful criteria, applied to the rift-basin deposits of Greenland, for distinguishing subaqueous burrows from subaerial ones. Homogeneous filling material, indistinct burrow walls, and the general absence of striae on burrow walls suggest soft subaqueous sediments, while striae and "vuggy" (containing open pockets, often filled with sparry calcite) burrow fillings indicate subaerial conditions.

Other evidence can be gleaned from sources that are not strictly sedimentological in nature. Not all lake basins contain basalt flows, but when present, the information they provide can be significant. The lateral extent of pillowed zones in basalts may be used as an indication of the dimensions of lakes, while the thickness of the zone (if it is only part of a lava-flow unit) may suggest water depths. The rates of the thinning of basalt flows as they pinch out against alluvial fans may eventually provide an indication of the slope of the fan. Bain (1941) reported that the Holyoke Basalt in the Hartford basin thinned from 170 meters to 30 meters (over 3.2 kilometers) as it lapped onto the eastern basin-margin alluvial-fan conglomerates. This would suggest an average local fan slope of 2.5 degrees if the top surface of the basalt flows can be assumed to be horizontal. This, combined with the position of shoreline deposits on the fan, might be used to estimate depths of lake waters.

Lacustrine deposits, in turn, can be used as evidence to solve other geological problems. The deep-water/offshore strata are laterally extensive compared to the other facies in the rift basins, and their thickness variation (or lack of it) can be used to infer relative rates of subsidence across the basin. Examples have been given of lacustrine deposits that thicken toward the major border faults of the basin. Additionally, Olsen (1985a) has suggested that the apparently uniform thickness of the lake beds across major intrabasin faults in the Newark basin indicates that these faults were not active during deposition of the lacustrine sequences.

9 / Playa Deposits

Modern playa environments have been known to American geologists ever since the government-sponsored exploration expeditions to the western United States in the late 1880s, but they have only recently been recognized in the Triassic–Jurassic rift-basin sedimentary record. Van Houten (1964, 1965*a*, 1965*b*) suggested that playas predominated during the repeated desiccation stages of the Lockatong lake, but most studies that have used the playa environmental model have been done since the late 1970s.

Interestingly, this environmental model has been useful almost exclusively in applications to the North American rift basins. In Morocco and Europe, most of the mudstone sequences are fundamentally different, extending widely beyond the limits of the fault-bounded graben depocenters and containing locally significant accumulations of evaporite minerals. Many of these sequences are probably marine-influenced, broad, sabkha environments; these are covered in chapter 10.

Playa (the term is Spanish for beach) usually refers to the dry flat area that remains after evaporation of a shallow ephemeral lake in an arid to semiarid climate. The sediments that compose this flat lake bottom may range from predominantly chemical to predominantly clastic. Recent studies have shown that thick halite deposits ("hundreds of meters thick," according to Lowenstein and Hardie 1985) have formed in nonmarine playa environments, but evaporites are only of minor importance in the Triassic–Jurassic rift-basin playas.

The length of time during which the basin actually contains water (a playa lake) and the frequency of filling are also highly variable; thus the diagnostic sedimentary structures range from undisturbed lacustrine features to secondary features created by soil-forming processes. Clemmensen (1979) has even applied the term *playa* to Triassic deposits in Greenland that were interpreted as originating in a nonsaline lake that, although shallow and ephemeral, was often stable enough to have supported well-developed stromatolites.

Primary Playa Deposits

The playas present in the North American rift basins were located in the low-lying areas of fault-bounded basins; they were similar in many respects to the modern Basin and Range playas of the southwestern United States. Perhaps the most extensive and best studied playa deposits are those of the Blomidon Formation in Nova Scotia. Klein (1962b) had interpreted these sediments as lacustrine in origin, but Hubert and Hyde (1982) have reinterpreted most of this formation, concluding that about three-fourths of it is of playa-mudflat origin.

Hubert and Hyde (1982) envisioned the playa mudflats as having been deposited from "mud-laden waters", and as being distally equivalent to the fan-toe sandy sheetflood deposits. The principal characteristics of these mudstones include the following: 10-centimeter- to 3-meter-thick bedding units of uniform thickness over hundreds of meters laterally (figure 9-1); very poor sorting of the sands, silts, and clays within the beds; the presence of mud cracks, carbonate nodules, traces of evaporite minerals; and an absence of invertebrate fossils or root traces. The presence of scattered eolian sand grains was also noted, supporting an interpretation of adhesion ripples for ill-defined, small-scale, lenticular bedding in some of the sandy layers (figure 9-2). Smoot and Olsen (1985) have since suggested that this type of bedding may be the result of pods of wind-blown sand that became trapped in small irregularities caused by (unpreserved) evaporite efflorescence on the playa surface.

Mud-flat mudstones are about six times more common than the associated playa-lake claystones in the Blomidon Formation. The claystones are distinctly different deposits. They are well sorted and fissile, containing several kinds of fresh-water fossils (including conchostracans and fish scales), while lacking or rarely containing mud cracks, lenticular bedding, or evaporites. The claystones are also of uniform thickness for hundreds of meters laterally.

Other playa environments have been inferred from deposits in the Hartford basin (Hubert, Gilchrist, and Reed 1982; Gierlowski-Kordesch 1985), the Central East Greenland basin (Clemmensen 1978a), the Newark basin (Van Houten 1964, 1965a, 1965b, 1969), and in South Wales (Tucker 1977). These interpretations are based on laterally persistent, poorly sorted mudstone beds of uniform thickness, and on the intermixed association of subaqueous sedimentary structures (such as oscillation ripple marks) with subaerial features (such as mud cracks, raindrop imprints, and vertebrate

Figure 9-1. Laterally extensive bedding and eight fining-upward graded sandstone-to-mudstone packages: playa deposits in the Fundy basin, Canada. Light-colored bands are graded sandstones; dark-colored bands are mudstones. Also note the two channel sandstones (Sc), present in this outcrop but rare elsewhere. Man for scale (arrow). (Photo courtesy of J. F. Hubert, from Hubert and Hyde 1982, Figure 3B. Reprinted with the permission of *Sedimentology***)**

tracks). Also noted by Van Houten (1965*a*, 1969) is the common absence of associated fluvial channels and the common presence of casts of probable glauberite crystals.

Wheeler and Textoris (1978) described a unique type of playa deposit in the Deep River basin of North Carolina. They suggested that thin beds of limestone and chert in this basin were deposited inorganically in playa lakes. The limestone occurs as 1- to 20-centimeter-thick beds of "limestone tufa . . . laminated, micritic, nonporous, and pelletoidal" (p. 768) and contains spar-filled voids. Because no algal filaments were found, the limestone was interpreted as an inorganic precipitate that encrusted plants growing in the lake.

The chert beds, up to 60 centimeters thick, are "nearly pure" crystalline quartz. In the absence of associated siliceous fossils, ghosts of siliceous fossils, or volcanic ash beds—and because, petrographically, the chert proved to be length-fast, rather than the length-slow variety commonly associated with evaporites—Wheeler and Textoris suggested that the chert originated as an inorganic opaline gel on the floor of the playa lake. Such deposits are common in the alkaline waters of modern rift-basin lakes in East Africa.

Figure 9-2. Wavy or lenticular laminations of sandy mudstone within mudstones, attributed to adhesion ripples in a playa environment in the Fundy basin. Coin for scale. (Photo courtesy of J. F. Hubert, from Hubert and Hyde 1982, Figure 3C. Reprinted with the permission of *Sedimentology*)

Trace fossils provide some evidence of ephemeral and unstable subaqueous environments. Gierlowski-Kordesch (1985) noted a number of escape burrows in the East Berlin Formation of the Hartford basin, inferring irregular events of rapid sedimentation in a playa environment. Gierlowski-Kordesch also attributed the small size of the trace fossils to the limited time available for biological development during the short, ephemeral high-water stages of the lakes. Boyer (1979) reported a distinctive trace fossil in the Brunswick Formation of the Newark basin that showed rapid rest and crawl alternations. This trace—associated with mud cracks, crystal casts of possible halite, and vertebrate tracks—was interpreted as being the record of an organism trying desperately to find a comfortable position in the increasingly inhospitable environment of a shrinking pool of water.

Secondary Modifications to Playa Deposits

Playa deposits are subject to repeated episodes of flooding and desiccation and to soil-forming processes during long periods of subaerial exposure. Thus, like those of floodplain deposits, the diagnostic characteristics of playa deposits are often the result of secondary modifications rather than the primary depositional sedimentary structures.

Van Houten (1962, 1964, 1980) discussed most of the secondary features superimposed on the late-stage Lockatong lake-cycle deposits. He suggested that the lakes had become "carbonate-clay playas with salts crystallizing in the mud, and repeated wetting and drying producing extensive cracking, brecciation, and upward-concave patterns [figure 9-3] of shearing (gilgai)" (Van Houten 1980, p. 269). Blodgett (1985) has recently described the development of

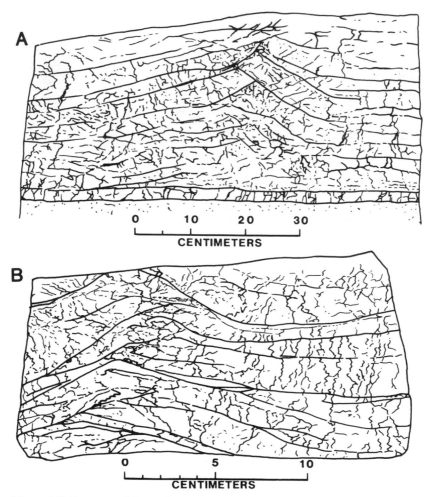

Figure 9-3. Patterns of disruption in the playa mudstones of the Lockatong Formation, Newark basin, New Jersey. Pattern (B) shows the central area of (A) in detail. (From Van Houten 1964, Figure 14. Reprinted with the permission of the Kansas Geological Survey)

similar upward-concave patterns of shearing in vertisols common to strongly seasonal or semiarid climates, where the soils are subject to multiple episodes of wetting and drying; this supports Van Houten's conclusions.

The extensive cracking and brecciation of the mudstones, commonly yielding a "shattered" texture of microscopic fragments, was originally ascribed to postdepositional burial and dewatering (Van Houten 1962). More recently, it has been interpreted as a "crumb fabric"—the ultimate result of the repeated wet and dry cycles common to the playa environment (Van Houten 1980; Smoot and Olsen 1985). Smoot and Olsen (1985, p. 30) have described this facies as millimeter-scale "mud clumps and abundant laminoid and ovoid cement-filled vesicles" with common clay linings. Such disrupted-mudstone fabrics (figure 9-4) have also been recognized in the Hartford basin (Demicco and Gierlowski-Kordesch 1986) and have been used to support an interpretation of a playa depositional environment there.

In other playa settings, thin layers of mud settled out of the shallow, muddy, stagnant waters of playa lakes and puddles. On desiccation, mud

Figure 9-4. Disrupted mudstones, produced by repeated episodes of wetting and desiccation of muddy sediments in a playa environment, Hartford basin, Connecticut. This stage is a precursor to the chaotic microbrecciation or crumb fabrics of some playa deposits. Similar rock fabrics were noted from the Lockatong formation in Van Houten's pioneering work (1964, Figure 13B). (Photo by the author)

curls formed as the mud cracked and contracted differentially. Such curls are usually ripped up and incorporated into fluvial sediments as intraclasts, but they can be preserved in playa environments where wind-blown sand fills the voids beneath the in situ curls and eventually buries them without destroying them (figure 9-5). Sand-filled deeper mud cracks are also common in this environment (figure 9-6).

Facies: Patterns and Controls

Playa environments record arid conditions and therefore are not usually associated with fluvial deposits. Rather, playa sediments are primarily facies equivalent to (and therefore interbedded with) dry-climate alluvial-fan deposits. They are most intimately associated with distal-fan sheetflood/sand-flat deposits. Playa deposits are also found in association with the rarer and more randomly distributed eolian deposits. They may occur interbedded with the deposits of more stable, perennial lakes, in which case the perennial lake deposits may grade upward into playa deposits (as in the Lockatong lacustrine cycles described in chapter 8). This sequence probably reflects a regional variation in climate rather than lateral shifts of facies.

Summary

Deposits of playa environments contain both primary and secondary sedimentary structures, but these environments are new to the suite of possible sedimentary interpretations in the rift basins, and few detailed studies have yet been made using this model. Van Houten did exceptionally well in his interpretations of playa-lake sedimentary characteristics, despite the fact that his interpretations were made prior to most sedimentological studies of modern playa environments. As is the case with most environmental models, however, studies of the modern analogues have produced considerably more detail than has yet been definitively recognized in the ancient sedimentary record.

Localized (basin-scale) playa deposits have thus far been documented only in North American and possibly in Greenland basins. They consist of primarily clastic deposits in these basins. Although traces of associated halite and nonmarine evaporite minerals are present, they do not occur in thick, monomineralic beds. The existence of playa deposits reflects the predominantly arid climate of the era and location, as well as the tectonically produced basins of internal drainage. The mainly clastic nature of the deposits, however, records the composition and relief of the source area.

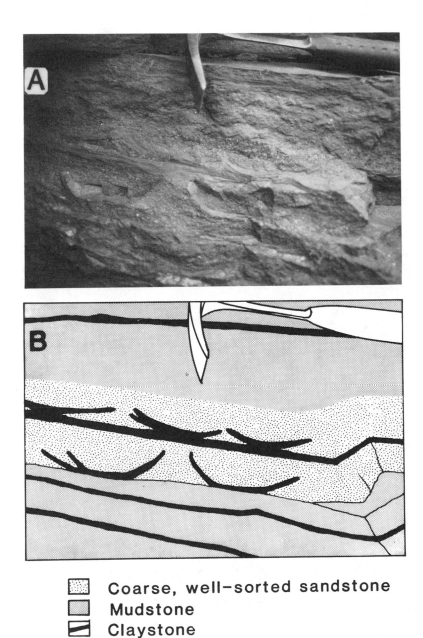

	Coarse, well-sorted sandstone
	Mudstone
	Claystone

Figure 9-5. Desiccated mud curls, preserved by infilling and burial by coarse, well-sorted, wind-blown sand, Hartford basin, Connecticut. (A) Photograph. (B) Tracing from photograph. (Photo by the author)

Figure 9-6. (A) Well-sorted, coarse-grained, wind-blown sand, filling large desiccation crack in mudstones (lens cap for scale). (B) Sandstone that filled desiccation crack in unconsolidated mudstone. Postdepositional compaction of the mudstone contorted the sandstone dike. Both rocks are from playa deposits in the Hartford basin, Connecticut. (Both photos by the author)

Data Collection and Interpretation Techniques Illustrated by Studies of Rift-Basin Playa Deposits

Many sedimentological field-study techniques that have proved useful are described in the studies mentioned in this chapter. The most important of these, as always, is the recognition of the significance of sedimentary structures, both primary and secondary. Mud cracks, which are ubiquitous in playa deposits, were for years accorded a significance beyond that of indicating subaerial desiccation: in the Hartford basin in the 1800s, they were used as definitive evidence of desiccation on a tidally exposed estuarine shoreline. At the same time, the associated crumb fabric, which appears to be environmentally diagnostic on a secondary level, was ignored because it was not understood.

Other important sedimentary features that have been described in these rift-basin deposits include the poorly sorted nature of the constant-thickness mudstone beds, and the juxtaposition of subaerial and subaqueous sedimentary structures. The paucity of rooting and burrowing has been implicit in these descriptions.

More recently, recognition of a distinction between adhesion ripples (which are hard to define and document in the ancient) and the unnamed "sand-silt pods" of Smoot and Olsen (1985) illustrates a finer level of environmental interpretation. The first feature is produced as damp sand or mud traps wind-blown sediment by surface tension; the second occurs when deeper saline groundwaters are drawn to the surface by capillary action and evaporated, and the irregularities in the resulting evaporite efflorescence mechanically trap wind-blown sediment.

The presence and uncommon composition of associated evaporite minerals, such as thenardite and glauberite, offer a key to the recognition of evaporitic environments and to the interpretation of the nonmarine nature of the brines from which the minerals precipitated. Equally, the associations of the playa deposits with other nonmarine deposits, such as "dry" alluvial fans and eolian sediments, are important in distinguishing playa from shallow, evaporitic marine environments. The absence of sandy fluvial channels and rooted floodplain mudstones is a strong indication of the nonfluvial character of such sediments.

The Triassic deposits in East Greenland also contain evaporites, although not in the vast volumes that the marine-influenced deposits discussed in chapter 10 do. Clemmensen (1978a, 1980) interpreted these as nonmarine evaporites on sedimentological evidence. Additional isotope studies reported by Clemmensen, Holser, and Winter (1984) support this interpretation. The sulfur $\delta^{34}S$ content of the sulfate evaporites has proved to be generally lighter than that of contemporaneous Triassic evaporites that are known to be marine. The values for both $\delta^{13}C$ and $\delta^{34}S$ from the Greenland formations were generally "low and scattered," which was taken to indicate the variable isotopic content of nonmarine source waters and the erratic rates of influx of these waters into the basin. Thus the Triassic evaporitic mudstones of East Greenland have been assigned primarily to a playa environment.

Finally, trace fossils have provided evidence of the episodic nature of deposition in playa environments. The significance and interpretations of trace fossils were often originally based as much on the characteristics of the sediments in which they were found as on the nature of the traces themselves, however, and care must be taken to avoid circular reasoning in this regard.

10 / Marine-influenced Deposits

Many of the Triassic rift basins in Europe and North Africa contain thick sequences of bedded halite and gypsiferous mudstone. These lithologies are part of a widespread evaporitic red mudstone facies that covered much of Europe and North Africa. It is significantly thicker—and locally, contains later-stage evaporites—in the grabens that were subsiding contemporaneously with deposition.

This facies was apparently related to an arid climate combined with the Late Triassic–Early Jurassic westward transgression of the Tethys over these low-relief continental areas (probably due to continental subsidence rather than to eustatic sea-level rise: Hallam 1984). Some areas may also have been flooded from the direction of the newly formed Atlantic Ocean and Arctic Ocean.

In northern Italy, adjacent to Tethys, nonmarine through deep-marine depositional environments existed in rift basins throughout Triassic–Early Jurassic time. These rift basins, created by tectonic adjustments of the Mediterranean microplates, contain sequences of deposits that grade upward from basal subaerial alluvial-fan deposits, through fluvial strata, into shallow-marine carbonates, and eventually to deep-marine turbidites and debris flows (Martini, Rau, and Tongiorgi 1986). These basins will not be discussed further, since the emphasis here is primarily on the nonmarine deposits along the developing Atlantic seaway, but they provide excellent examples of sedimentation controlled by tectonics, and of the effects of structural complexities within the basins on patterns of sedimentation.

Fully marine conditions were rare in the basins of northern Europe and northwestern Africa, and the exact nature of most of these deposits has been and is still the subject of discussion. The proposed environments have included lakes, inland seas, shallow-marine systems, vast continental playas, and vast coastal sabkhas. The basic controversies are over the questions of water depth, sources of the waters and of the brine concentrations, and the frequency with which "full" water depth was present.

A distinction can and should be made between the thick halite beds (which onshore are confined to the graben areas) and the sulfate-bearing red mudstones that are ubiquitous. The two types of evaporites often formed synchronous facies within the same regional setting, but different depositional environments and different lithologies were produced by the local variations in tectonics. Both facies have often been interpreted in the context of a single depositional model, causing some confusion in the literature as well as difficulty in separating the sedimentary concepts illustrated by the distinctly different environments. Therefore, the salina, sabkha, and intertidal subenvironments have not been separated here as they would have been in the other chapters, and a more basin-specific type of discussion is presented.

The halites are thick enough in several places in Morocco to have formed salt diapirs both onshore and offshore. Halite deposits several kilometers thick are also present offshore in the western North Atlantic, where they are apparently not confined to grabens, and where they are inferred to record flooding of the opening North Atlantic seaway by Tethyian waters (Jansa, Bujak, and Williams 1980). This evaporitic facies is not present in the onshore North American rift basins, which were farther from the sources of marine waters.

The widespread evaporitic deposits owe their origin to a combination of three factors: low-relief topography, which produced only muddy sediments during erosion (figure 10-1); an arid climate in which evaporation of water and concentration of brines could take place; and a tectonic and eustatic setting in which sea level and the general level of the low-relief continents were usually very close to each other. More rapidly subsiding grabens caused local variations in this general setting and led to the deposition of thick halite beds within the basins.

The following descriptions concentrate on the better-known deposits of Morocco and England, where these evaporitic mudstones are strongly influenced by local syndepositional graben formation. Similar sequences occur in France and on the Iberian peninsula, but they are not as well known.

Because of the solubility of the evaporites and the fine-grained nature of the associated clastics, this facies is poorly exposed; consequently, this chapter also discusses the subsurface data that have been used in interpretations. Many of the halite deposits were in fact unknown until the beginnings of subsurface economic exploration of the regions.

Figure 10-1. Exhumed low-relief Triassic topography that was subject to
inundation, Kerrouchen area, Morocco. Hills in the background are composed
of red Triassic mudstones. Light-colored rocks in the foreground and middle
distances are smooth, eroded remnants of the granite that had been the source
terrane for alluvial-fan conglomerates in the nearby graben. Pattison, Smith,
and Warrington (1973) described a similar terrane in England as probably
having been a "paleogeographic vista of almost unrivaled monotony." (Photo
by the author)

Moroccan Salt Basins

Many of the early-twentieth-century French geologists who worked in Morocco
viewed the widespread evaporitic mudstone facies of the Moroccan Triassic
as "continental" (nonmarine) deposits. In 1962, however, Lucas published a
detailed study of the clay mineralogy of this facies. He concluded that the
mineralogy of the "neoformed" (authigenic) clays in many areas of Morocco
was indicative of an environment of restricted, shallow, "continental" seas
possessing a geochemistry similar to that of an alkaline lake. The conclusion
was tentative and ambiguously supported, but it provided a platform for
geologists to use in extending the existing marine interpretations of the
known halite-bearing Algerian Triassic formations westward into Morocco.
Simultaneously, subsurface exploration provided wells that showed the
presence of thick halite beds in Morocco (figure 10-2). Some of the halites in

Figure 10-2. Thick deposit of bedded salt (about 20 meters shown) within evaporitic mudstones, Central-N'Fis basin, Morocco. This artificial exposure was created by surface hydraulic/solution mining operations. (Photo courtesy of F. B. Van Houten)

the central coastal plain regions of Morocco contain late-stage evaporite minerals (Amade 1965), further supporting the marine interpretations.

Salvan (1968a, 1968b, 1974a, 1974b, 1983) is the major proponent of the marine model for the Moroccan evaporites. In addition to Lucas's clay study, Salvan's principal evidence is what appears to be a lateral fractionation series of minerals across Morocco, a series that could have been precipitated from increasingly concentrated and depleted marine brines. This apparent series of mineral phases begins with shallow-marine Alpine-type Triassic carbonates in the Rif Mountains and progresses southwestward through sulfates, chlorides, and finally to the local end-member potash minerals (figure 10-3). It, together with locally occurring marine fossils in the carbonates in easternmost Morocco, led Salvan to postulate that the environment was a vast continental platform of generally shallow water ("more or less deep") of marine origin, with irregular influxes of new seawater and dilution of the brines. Freshwater/nonmarine sedimentation took place near the highstanding Atlas Mountains, while deep-marine waters were inferred to have

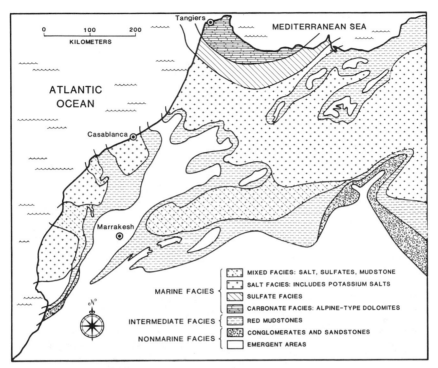

Figure 10-3. Distribution of Triassic evaporite and evaporitic mudstone facies across Morocco. (After Salvan, 1974*b*, Figure 2. Reproduced with the permission of the Société Géologique de France)

deposited thick halites and potash minerals in the deep basins formed by the grabens of the coastal plain area.

Since Salvan's early studies, the Rif Mountains containing the carbonates of this inferred fractionation pattern have been interpreted as an allochthonous tectonic terrane (Bosellini and Hsü 1973). Moreover, the pattern of evaporite mineral phases across Morocco is much more erratic than that of a uniform fractionation series (figure 10-3). Although Salvan's premise that the area was subject to episodic marine incursions has not been challenged, Van Houten (1977, p. 85) suggested instead that the thick halites in the grabens represent shallower "structurally restricted saline lagoons fed by marginal marine incursions," and that the more widespread evaporitic mudstones were deposited on vast salt pans and mud flats that were subaerially exposed most of the time rather than lying under shallow but always-present marine waters.

Unfortunately, except for brief descriptions of oscillation ripples and herring-bone crossbedding (Petit and Beauchamp 1986), the bedding characteristics and sedimentary structures that might support or disprove these environmental interpretations have not been published. Some of the more resistant beds within the evaporitic mudstones have a high carbonate content— probably dolomite (figure 10-4)—that suggests concentrated, magnesium-rich brines. This type of lithologic fabric has been briefly described in the similar Triassic evaporitic mudstones of England by Taylor (1983) and by Jeans (1978) in sequences interpreted as marine-influenced evaporitic facies, but this in itself does not definitively point to either marine or playa conditions.

Van Houten's interpretation of the widespread mudstones has elicited discussion but little in the way of new sedimentological evidence. Beauchamp (1983) and Petit and Beauchamp (1986) have published a map of crossbed vectors that are interpreted to be littoral longshore currents in the Central-N'Fis basin (figure 10-5), an inland graben complex that does not contain

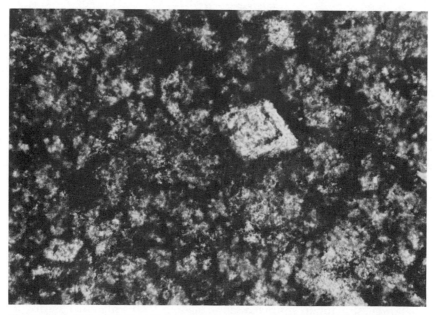

Figure 10-4. Dolomite rhombs in a red carbonate–clay lithology that forms resistant ledges within otherwise soft evaporitic mudstones in the Kerrouchen basin, Morocco. Width of the area shown in the photo is about 1.5 millimeters. Presence of dolomite indicates the presence of magnesium-rich brines at some point during or shortly after deposition. (Photo by the author)

abundant thick halites. These vectors are diametrically opposed to those measured by Mattis (1977), who inferred a fluvial environment for these rocks (compare figure 10-5 to figure 5-9). The orientation of these paleocurrents is critical not only to interpretations of environment, but also to whether the Massif Ancien, between the Central-N'Fis and Essaouira basins, was a high-standing Triassic source area (Jones 1975; Mattis 1977; Brown 1980), or whether it was uplifted later to bisect what had been a continuous, south-westerly, basin-axis Triassic drainage pattern along the two basins (Petit and Beauchamp 1986).

Biron (1982) has suggested that there may be turbidite deposits in this same basin—although, as diagramed, they more resemble point-bar deposits in scale and structures. Some pelecypod fossils (*Pholadomya* sp.) and possibly

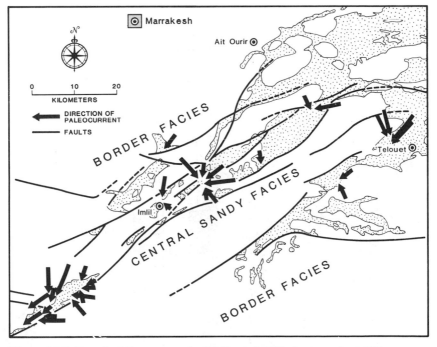

Figure 10-5. Crossbedding vectors in the Central-N'Fis basin region of Morocco, as measured by Beauchamp (1983, Figure 6). The number of measurements was not indicated, but the length of the arrows is proportional to the consistency ratio of the crossbed measurements at that locality (Petit and Beauchamp 1986). Beauchamp suggested that paleoflow southwestward down the basin was caused by longshore currents along the graben axis. (Reproduced with the permission of the Faculté des Sciences, Université Cadi Ayyad)

fragments of echinoderms and brachiopods have been reported from this sequence (Biron 1982; Beauchamp 1983; Petit and Beauchamp 1986). Detailed descriptions of these fossils have not yet been published, but they have been used to support the shallow-marine (lagoons, beaches, tidal channels, and tidal flats) interpretations.

The extent and thickness of the pillowed zones in the several basalt flows intercalated within the evaporitic mudstones in Morocco may be an important piece of evidence. These provide data for interpretations of how frequently the continental platform was inundated and how deep the waters were when they were present. Several studies have shown that the basalts associated with the evaporitic red mudstones are not normally pillowed. Where they are, the pillowed zones extend only hundreds of meters laterally and are only a few tens of meters thick (figure 10-6). The tops of the same flows often display subaerial columnar jointing (Cogney, Termier, and Termier 1971; Cogney et al. 1974; Mattis 1977; Cogney and Faugères 1975).

Thus, the basalt flows suggest that water was present only in local ponded areas on mud flats at the time of lava extrusion. The extent of pillowing of

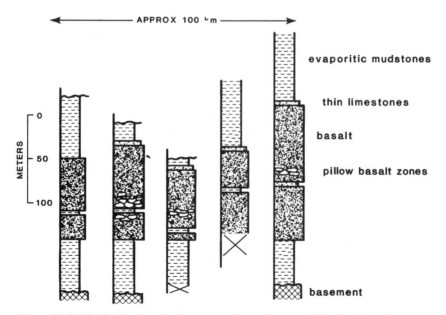

Figure 10-6. The limited vertical extent and erratic lateral distribution of pillowed zones within the ubiquitous basalt flows (from the thin zones of sedimentation between grabens in Morocco) suggest that marine incursions and standing water were infrequent in these areas. (After Cogney and Faugères 1975, Figure 2. Reproduced with the permission of the Société Géologique de France)

the basalts where they directly overlie thick, bedded halites is not indicated in published sections (for example, in Salvan 1968*a*, Figure 13), and thus water depths in the grabens/salinas cannot be inferred. In contrast to the thick, extensive pillowed lavas that are found in some of the Italian rifts in association with obviously marine deposits such as turbidites and submarine fan breccias (Martini, Rau, and Tongiorgi 1986), the locally pillowed or unpillowed Moroccan basalts argue against deep-marine and/or constantly marine conditions.

The basalt flows also provide evidence on the relative rates of subsidence and sedimentary infilling in the grabens and the intergraben areas. Salvan (1974*b*) postulated that the grabens were essentially deep, subaqueous, clastic-starved basins, in order to apply a deep-water evaporite precipitation model and to account for the potash minerals and thick halites. Since the basalt flows do not change thickness significantly across the grabens, however, Salvan (1983) also had to postulate rapid salt deposition, such that at the time of basalt extrusion, the level of the depositional surface in the grabens was essentially equivalent to that in adjacent areas. Synchronous subsidence and sedimentary infilling in a shallower environment would eliminate the necessity for this fortuitous sequence of circumstances, especially inasmuch as the basalt flows do not mark the end of halite and evaporitic mudstone deposition.

The virtual absence of siliciclastic material other than clay in much of the graben-fill salt deposits also suggests that, in these instances, graben-floor subsidence occurred rather than basin-margin elevation or some combination of both. Sedimentation probably kept the depositional floor of the graben near the surface of the adjacent horsts, and there was little topographic relief between them, and therefore no rapid erosion and no coarse clastic material resulted.

Amade (1965) has provided the only published detailed description of the thick Moroccan halite sequences to date. This was a study of the subsurface section in the fault-bounded Khémisset basin in north-central Morocco. In this basin, over 1,000 meters of sediment was deposited, much of it halite in beds with less than a half-percent clay content. The salt beds pinch out at the borders of the basin, and the different evaporite minerals form a rough, concentric, fractionation pattern (figure 10-7); the most soluble minerals (sylvite, carnallite, rinneite) are found in the central parts of the basin. They are surrounded by halite, while sulfates (gypsum and anhydrite) occur along the margins of the basin.

Amade was aware of the theoretical sequence of precipitation of minerals from evaporating sea water; and upon finding sylvite deposits surrounding and overlying a thick carnallite bed in an apparent inversion of the experimentally determined sequence, he hypothesized that there had prob-

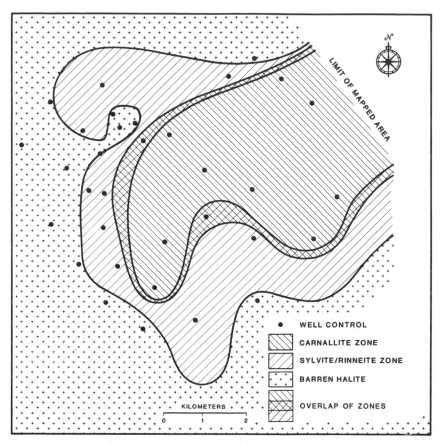

Figure 10-7. Subsurface concentric evaporite patterns in the Khémisset basin, Morocco. The halite, sylvite, and carnallite beds are arranged in order of increasing solubility toward the center. Moreover, within each mineral zone, the thickness and concentration of the mineral increases toward the center of the basin. Details of the shape of the pattern shown here may be more a function of the distribution of the available control points than of anything else, but the localization of the potash minerals and the generally concentric pattern are valid. (After Amade 1965, Figure 5. Reproduced with the permission of the Direction de la Géologie de Maroc)

ably been postdepositional dissolution and reprecipitation of parts of the mineral sequence. Postdepositional and syndepositional recrystallization and recombination of mineral phases is common in the relatively unstable evaporite minerals, and sylvite is commonly formed as an incongruent dissolution product of carnallite in undersaturated waters (Hardie 1984). There are also numerous isolated sylvite beds within the halites of the Khémisset basin, reflecting the complex internal horst and graben structures in the basin. Amade suggested that the differing mineralogical sequences of these isolated deposits indicated that each subbasin had undergone an individual history of brine evolution.

The large volume of evaporite minerals, and the absence of typically nonmarine evaporite minerals (such as the glauberite and thenardite found in the North American playa deposits), suggest a marine origin for the brines from which the evaporites precipitated. On the other hand, the heterogeneous nature of the different deposits suggests that the environment was not a homogeneous, deep-marine basin, but rather more akin to Van Houten's (1977) concept of isolated, episodically inundated saline lagoons and salt flats. The localization of the end-member evaporites toward the axes of the basins suggests that there was some relief to the basin floors, and that with successive inundations the more soluble components were preferentially dissolved and washed into the central low-lying areas, where they concentrated. This could have occurred subaerially, subaqueously, or in an environment that alternated between the two.

In the Essaouira basin of southwestern Morocco, Brown (1980) and Harding and Brown (1975) have documented a slightly different suite of marine-influenced deposits. A nonmarine to marine vertical and lateral facies sequence was deposited there as waters from the newly formed Atlantic Ocean flooded eastward into the basin, leaving a record of vertical and westward gradation from fluvial to evaporitic sabkha-flat (figure 10-8) to shallow-marine environments. The transgression was generally (but not entirely) restricted to the area of the graben. As elsewhere, the deposits are thicker in the graben, but these evaporitic, sabkha mudstones spread beyond the graben borders.

English Salt Basins

The widespread, evaporite-bearing red mudstone facies of the English Mercia Mudstone Group (formerly the Keuper Marl) has also been the subject of a number of different environmental interpretations. These interpretations have ranged from marine to paralic to inland-lake or playa (Arthurton 1980). Like the Moroccan deposits, the thick salt deposits are localized within fault-bounded grabens, whereas sulfates are common in the more extensive mudstones. Unlike the situation in most of Morocco, however, there is often a distinctive transitional formation that marks the grada-

Figure 10-8. Irregular beds of gypsum, anhydrite, and red mudstone of probable sabkha origin, Essaouira basin, Morocco. (Photo courtesy of R. H. Brown)

tion from underlying fluvial graben-fill sediments up into the evaporites and evaporitic mudstones. As in Morocco, the thick halites of the grabens are known only from subsurface studies.

Origin of the Brines

There are presently few adherents to theories of deep-water marine or open-ocean origin for the English Triassic evaporites, but evidence suggests that the brines from which they precipitated did have a marine source. The associated transitional facies between the nonmarine deposits and the evaporite-bearing sediments contain marine to brackish-water bivalves (*Lingula*, Rose and Kent 1955) and a limited assemblage of acritarchs, interpreted as marine microplankton (Warrington 1967). The fossil assemblage is not as varied as it would have been in an open-marine community; rather, it is of limited variety, indicating a stressed environment. Warrington (1967) inferred a near-shore or estuarine environment from this low-diversity suite of fossils.

Ireland et al. (1978) have interpreted the sedimentary structures of this transitional facies in parts of England as representing intertidal mud and sand flats, supratidal sabkhas, and rare tidal channels. Thus, the limited sedimentary and fossil evidence suggests that the environments just prior to deposition of the thick halites in the grabens were becoming more marine, transitionally from the older, predominantly fluvial strata.

Sedimentary structures and fossils within the mudstones and halites themselves, however, are rare. Although Evans et al. (1968) suggested that the "sheer volume" of halite in the English rift basins—as well as the general absence of siliciclastic sediments in the halite sequences—requires a marine source for the brines, direct evidence from the halites as to their origin is primarily geochemical in nature.

The bromine content of graben-fill halites reported by Haslam, Allberry, and Moses (1950) was cited by Tucker and Tucker (1981) and by Holser and Wilgus (1981) as suggestive of marine source waters. The element bromine forms a solid solution series with chlorine in halite, but during salt precipitation a smaller ratio of bromine to chlorine is always incorporated into the halite crystals than existed in the brine. Thus, while typical primary marine halites contain approximately 40 to 200 ppm bromine (Holser and Wilgus 1981), with each re-solution and reprecipitation event these salts will (in theory) contain an increasingly smaller bromine content. Nonmarine halites typically contain about 1 to 20 ppm bromine.

The bromine analyses of the English halites are within the range of primary marine halites. More significantly, they display a "comparatively small variation" (Haslam, Allberry, and Moses 1950), consistent with a homogeneous marine source rather than with heterogeneous values common in deposits of more variable recycled-evaporite or entirely nonmarine sources.

Clastic deposits from areas outside of the grabens contain clues as to the origin of the brines. Klein (1962a) has suggested a marine influence based on sedimentological structures from within the Mercia Mudstone Group. He suggested that this is a marine sequence, based on the presence of flat-topped oscillation ripples, and "air heave" structures. The features were inferred to be indicative of tidal flats and tidal fluctuations, where the crests of ripples were eroded at low tide and where air that had been entrained in sediments was forced out during the rising tides, disrupting the sediments.

Brines of marine origin are variously thought of as having been introduced by episodic marine flooding during eustatic fluctuations (Arthurton 1980) or as being the product of a more gradual migration of increasingly concentrated and mineralogically depleted brines across vast continental platforms (Warrington 1974). In the latter case, the brines are usually envisioned as having been a constant shallow-water presence, rather than as having been ephemeral with frequent episodes of desiccation. It has not yet been possible to support such speculations with conclusive data. Regional lateral fractionation patterns of minerals are vague at best, although on a local scale, the sulfate minerals are laterally equivalent to graben-center halites. Late-stage potassium-bearing minerals are unknown in England (Whittaker 1980), creating an as-yet unexplained mineral imbalance.

Several geologists have advocated a primarily nonmarine environment for parts of the Mercia Mudstone Group other than the thick halites. Tucker (1977, 1978) has suggested that both "subaerial (megaplaya/supratidal flat)" and "subaqueous (hypersaline lake/inland sea)" conditions existed at different times during deposition of this facies. As evidence for an environment with very little marine influence, Tucker cited an absence of marine fauna, structures indicative of periodic and extensive lake-shore regressions, the presence of caliches, zones of packed gypsum nodules, and other subaerial indicators. Tucker has suggested that much of the evaporitic mudstone facies was in fact deposited in "continental sabkha" environments—present as transitional facies between fluvial and hypersaline lake environments—and during lacustrine low stands. Tucker suggested that the homogeneous, red dolomitic mudstones are offshore lacustrine deposits.

Depths of Water

In addition to questions about the origin of the brines, there has been discussion as to the depth of water from which the graben-filling halites precipitated. Evans (1970) and Evans et al. (1968) have been the most recent proponents of deep-water, halite-precipitating environments in the grabens. They also suggest that shallow waters predominated over the surrounding "shelf" areas. Their evidence for deep water in the grabens included an apparent absence of desiccation phenomena, the absence of solution or erosion surfaces in the salts, and the absence of vertebrate tracks and eolian sediments. This negative evidence was taken to indicate subaqueous conditions "hundreds of feet" deep. The absence of late-stage evaporites was inferred to show that evaporation never reached completion, as it would in desert environments, while the thickness of the salt beds (exceeding 400 meters) was taken as an indication of the original, unfilled basin depths. As will be seen, most of this negative evidence has been superseded by newer, positive evidence that supports altogether different interpretations.

Other evidence cited by Evans (1970) for a deep-water origin for the halites is the lateral continuity of correlatable beds (figure 10-9)—although significant variability in the thicknesses of correlatable beds would seem to argue for syndepositional subsidence, rather than for simple infilling of an existing deep basin. Evans also suggested that the "four to five hundred" beds observed within 30 meters of halite may represent annual deposits; this he inferred to indicate deposition at a more rapid rate than could be accommo-

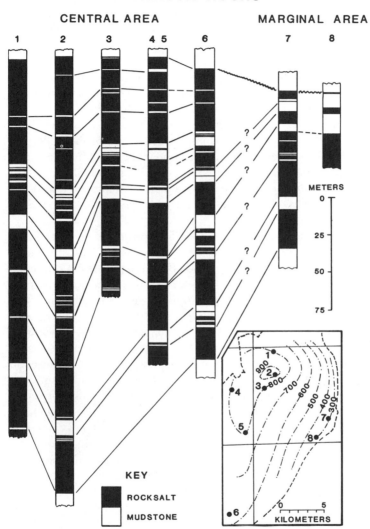

TRIASSIC ROCKS

CENTRAL AREA

MARGINAL AREA

KEY

■ ROCKSALT

□ MUDSTONE

Figure 10-9. Correlation of salt deposits within the Cheshire basin, England, showing good correlations of bedding despite lateral variability in thicknesses within the basin and suggesting deposition at one horizon (near sea level), rather than an originally deep, infilled basin. The correlation also shows abrupt thinning of halite deposits at the margin of the basin. Contours of the inset are isopachs (in feet) of the formation. (After Evans et al. 1968, Figure 23. Reproduced by permission of the director, British Geological Survey; U.K. crown rights reserved)

dated by subsidence, therefore necessitating an originally deep basin. There is no suggestion of a shallowing-upward sequence through this series of thin salt beds, however, and there is no firm evidence that the beds are indeed records of annual events.

Most geologists prefer models of a shallow-water to subaerial environment as appropriate for the halites and mudstones in England. Arthurton (1973, 1980) has reported both solution and eolian deflation surfaces in the halite deposits, as well as "salt-crust disruption structures"—all indicative of subaerial exposure. Arthurton (1973) has also done critical work on the detailed sedimentology of the halites. The primary crystal fabric of the halite has been preserved where syndepositional mudstone separates individual halite crystals. When these fabrics were compared with experimentally produced halite fabrics, the results suggested that the graben-fill halites grew as floor-nucleated crystals and as overgrowths on detrital halite within brine bodies less than a meter deep.

A study that has indicated frequent subaerial exposure of the thick halite beds is that of Tucker (1981) and Tucker and Tucker (1981). This study took advantage of the relatively large lateral and vertical exposures of the halite deposits found in subsurface mine workings to document very large polygonal structures in the salt beds. The polygons, 6 to 14 meters in diameter, with vertical crack dimensions of up to 6 meters (figure 10-10A), were interpreted as being the product of volume changes of the halites associated with annual temperature fluctuations. Smaller cracks that might have been associated with daily temperature fluctuations were inferred to be absent because the subaerial salt-flat surface consisted of "unconsolidated mush." Numerous layers of filling within the vertical cracks (figure 10-10B) were interpreted as representing different episodes of crack reactivation, and truncated cracks (figure 10-10C) were interpreted as recording dissolution planes caused by influxes of new, unsaturated brines.

Arthurton's study of cores from wells drilled into mudstones and associated thick halites of the central graben facies showed cyclic deposits of alternating laminated and blocky mudstones (figure 10-11), each cycle being about 5 meters thick. The lower part of the sequence displays laminations of well-sorted mudstone and siltstone (on the order of a few centimeters thick) that are extensive and are often graded, crossbedded, and rippled. Evaporite pseudomorphs and desiccation cracks are common. Arthurton has interpreted these features as having been produced in a low-energy, subaqueous, laterally uniform, primarily nonmarine environment—one that was subject to flooding, but that usually was emergent and undergoing desiccation. The overlying blocky mudstone facies consists of unsorted, homogeneous, silty mudstones with traces of anhydrite, dolomite, and halite; it is suggested to represent subaerial accumulations of wind-blown material.

Arthurton has inferred that these cycles are of eustatic origin—the laminated facies having been associated with marine transgressions, and the blocky facies with regressions. The evidence for this conclusion is somewhat ambiguous, and it is not made clear exactly how eustatic fluctuations could have controlled primarily nonmarine deposits. Arthurton suggests that true marine flooding is represented only by the halite deposits, which seem to have interrupted the deposition of the laminated-blocky couplets at irregular intervals.

The upper, blocky parts of the cycles may represent total disruption of a previously deposited laminated facies, instead of primary wind-blown accumulations (although the vertical accretion of wind-blown material is common on sabkhas). Arthurton indicates that the frequency of mud cracks and curl-ups increases upward in the laminated facies. If so, they would be similar to the cycles of increasing desiccation and crumb fabrics recognized in some of the North American rift-basin lacustrine and playa deposits. Even if the general process was similar, however, the brine chemistry and environmental setting would still have produced a salina-to-sabkha cycle rather than a lacustrine-to-playa cycle of the type found in North America.

Support for the interpretation of an alternately wet and dry sabkha or megaplaya type of environment for parts of the Mercia Mudstone Group may be present in Dumbleton and West's (1967) engineering description of some of the mudstones. Concerned with the rock properties of the mudstones, they briefly described the mudstones' texture as an "aggregated structure . . . cuboid fragments [about 1 to 2 centimeters in size] . . . which are themselves made up of fine aggregations. . . ." (1967, p. 1). This texture is reminiscent of the crumb fabrics of disrupted mudstones that have been described in playa deposits in the North American rift basins.

These data support Tucker's (1977) conclusions that parts of this evaporitic facies, including the halite deposits, were often subaerially exposed during deposition. It is difficult to apply a label to this environment, but in general it seems to have been (as in Morocco) one of vast, low-relief, vertically accreting sabkha-type mud flats that were episodically inundated by concentrated marine brines and then desiccated. Clastic muds and syndepositional sulfate evaporites were deposited in this setting. Within this broad environment, local graben subsidence created large, shallow depressions in which salinas formed.

Shallow waters were present more nearly continuously in the salinas because they were depressions created by continued subsidence—although they, too, were subject to desiccation. After flooding by marine-derived

Figure 10-10. Thermal contraction cracks in halite: subterranean exposures in mine workings from the Cheshire basin, England. (A) Large-scale view of the depth and width of cracks, downward-tapering shape, and upturned edges of polygon near paleosurface (man for scale). (B) View of vertical multiple banding within a crack (composed of both chemical and clastic deposits), indicating multiple episodes of crack activity. (C) Close-up of lower, steeply upturned, older polygon margin, planar dissolution surface, and younger polygon margin that is upturned less steeply. (Photos courtesy of M. E. Tucker)

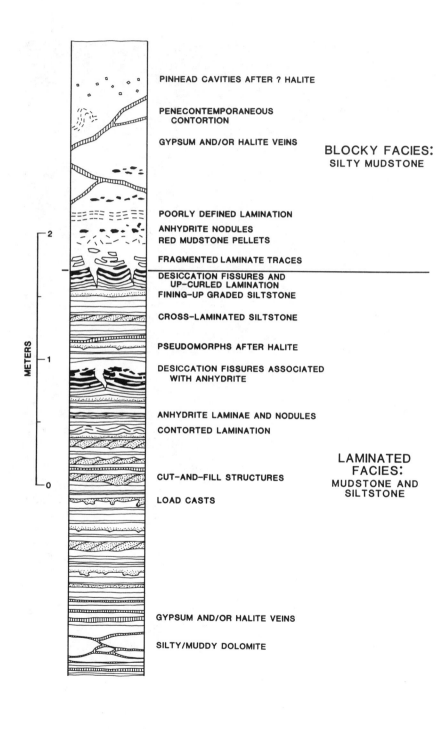

METERS

PINHEAD CAVITIES AFTER ? HALITE

PENECONTEMPORANEOUS
CONTORTION

GYPSUM AND/OR HALITE VEINS

BLOCKY FACIES:
SILTY MUDSTONE

POORLY DEFINED LAMINATION
ANHYDRITE NODULES
RED MUDSTONE PELLETS

FRAGMENTED LAMINATE TRACES

DESICCATION FISSURES AND
UP-CURLED LAMINATION
FINING-UP GRADED SILTSTONE

CROSS-LAMINATED SILTSTONE

PSEUDOMORPHS AFTER HALITE

DESICCATION FISSURES ASSOCIATED
WITH ANHYDRITE

ANHYDRITE LAMINAE AND NODULES
CONTORTED LAMINATION

**LAMINATED
FACIES:**
MUDSTONE AND
SILTSTONE

CUT-AND-FILL STRUCTURES

LOAD CASTS

GYPSUM AND/OR HALITE VEINS

SILTY/MUDDY DOLOMITE

2

1

0

brines, halite precipitated in these depressions and accumulated at a rate commensurate with subsidence. Rainstorms dissolved halites that had been precipitated in the surrounding mudstones; the waters then flowed into the salinas and increased their halite content while diminishing that of the mudstones. During episodes of flooding by less-concentrated brines, or perhaps by fresher waters from unspecified nonmarine sources, the sedimentation in the grabens consisted primarily of laminated clastic deposits that were similar to those of many lacustrine sequences.

Transgressive Facies

Several studies of the Triassic transgressive facies in England have been published, but it is not clear how closely the transgressive facies were bounded by the underlying rift basins. One of these studies (by Ireland et al. 1978) is presented briefly here, inasmuch as the transgressions occurred primarily in the regions of the rift basin and the younger deposits were probably somewhat limited in width by the same general tectonic framework (Whittaker 1980).

Ireland et al. (1978) have described tidally influenced deposits from the transitional facies between the underlying fluvial deposits and the previously discussed evaporite-bearing mudstones in central England. They reconstructed low-energy tidal-channel, intertidal sand- and mud-flat, and supratidal sabkha environments. The dominant lithology is a red muddy siltstone to silty mudstone, interpreted as "high intertidal mudflat to impersistent supratidal sabkha" (1978, p. 425) deposits. Although local horizontal laminations and ripple marks are preserved, the rocks are usually "structureless and ill-sorted" (p. 410). This evidence, in addition to common desiccation cracks, was used to infer alternately subaqueous and subaerial conditions, and either total bioturbation or a lack of variability in subaqueous flow conditions. Halite pseudomorphs and local beds of gypsum and dolomite were inferred to represent evaporite ponds in the sabkha surface.

Rare, isolated lenses and sheets of well-sorted sandstone (up to 2.5 meters thick) are present in the mudstones. Common intraclasts, clay drapes, internal scour surfaces, and reactivation surfaces were taken as evidence of variable (tidal) flow regimes. The grain sizes fine upward within a bed and,

Figure 10-11 *(facing page)*. **Schematic representation of one cycle of the cyclic deposits from the Triassic evaporitic mudstone facies of the Cheshire area, England. (After Arthurton 1980, Figure 3. Reproduced with the permission of the *Geological Journal*, copyright 1980, John Wiley & Sons)**

together with probable point-bar lateral accretion surfaces, suggest high sinuosity. Thus, the sandstones having a sheet morphology were inferred to record lateral migration of tidal channels.

Low intertidal sand flats have also been recognized by Ireland et al. Alternating tidal currents and slack waters were inferred to have produced 5-to 25-centimeter-thick beds of fine sandstone with internal crossbeds and ripples. Variable or tidal currents are also evidenced by flaser, wavy, and lenticular bedding, along with intraclasts and scours.

Higher on the paleoshoreline, high intertidal mud flats formed; in these, interbedded sandstones, siltstones, and mudstones were deposited as wispy, impersistent beds on the order of 5 centimeters thick. Desiccation cracks and halite pseudomorphs point to episodes of alternate subaerial and subaqueous conditions.

Ireland et al. have suggested that the broad, mixed sand and mud tidal flats of the North Sea may be good analogues to the reconstructed Triassic environments that deposited these predominantly muddy and evaporitic sediments. The general lack of clean sandstones was taken as indicative of the low energy of tidal and longshore currents and of waves, while the paucity of tidal channels pointed to both low tidal energy and low volumes of water being carried off the flats, in accordance with a generally arid climate.

Aquitaine Salt Basin

Salt-filled Triassic grabens occur in several other parts of Europe. One such graben, the last to be examined here, is the Aquitaine basin in southwestern France. As described by Stévaux (1971) and Stévaux and Winnock (1974), this subsurface basin has a compound structure that resulted in a variable pattern of detrital and evaporite mineral facies. In the central basin, a basal conglomeratic facies fines upward into the predominantly chemical phase of evaporites which is about 1,000 meters thick. From the basin margin toward the center, a succession of stratigraphically equivalent facies grades laterally from carbonates to sulfates to chlorides, with local occurrences of potash minerals within the chlorides (figure 10-12). A northern subbasin is filled predominantly with detrital sediments; it contains local sulfates and carbonates toward the basin center, but no salt.

Stévaux (1971) and Stévaux and Winnock (1974) have interpreted this basin-fill sequence as the record of increasingly reduced source-area relief (to account for the fining upward and eventual cessation of detrital input), combined with the flooding of a deep-water, silled basin from the Tethys to the south (and possibly from other marine waters to the east). The lower

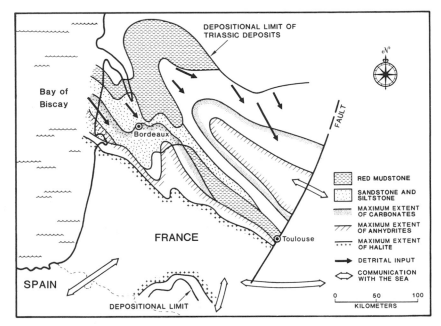

Figure 10-12. Distribution of detrital and evaporite minerals in the Aquitaine basin, southwestern France. Note the concentric zonation. (After Stévaux and Winnock 1974, Figure 8. Reproduced with the permission of the Société Géologique de France)

parts of the evaporitic sequence contain authigenic quartz and pyrite, suggesting episodically reducing and acidic confined environments during early stages of evaporite deposition. Such minerals may also be of diagenetic origin, however. The concentric pattern of evaporite facies is interpreted as being the result of complete dissociation of the brine from the marine sources, although the southern end of the basin is bounded by a post-Triassic fault and the continuation of the facies patterns to the south is unknown.

Basalts—probably flows, as they always occur in the same stratigraphic position—may eventually prove to be useful to the sedimentological interpretations of this basin. It is not obvious from the data presented whether the Aquitaine Triassic deposits are marine or nonmarine in origin; but by analogy to the thick halites in English and Moroccan basins, they are probably similarly marine-influenced. Completely circular facies patterns, if they could be definitively documented, may argue for nonmarine environments. The basin is situated, however, in an area over which marine brines are suggested to have traversed in order to reach the English Triassic deposits.

Marine Influence in the Central East Greenland Basin

A different type of possible marine influence has been noted in the rift basin in Greenland by Clemmensen (1976, 1980). Thick evaporites are not present in this basin, but the sedimentary structures in the clastics may indicate a tidal and/or marine influence during deposition of some sequences. The potential source of this marine influence is a narrow, extensive southern arm of the developing Arctic Ocean (Birkelund and Perch-Nielsen 1976). Clemmensen (1980) interpreted a sequence that contains rare marine bivalves and displays oscillation ripple marks and flaser bedding (along with desiccation cracks and mudstone rip-up clasts) as a brief marine transgression into an otherwise land-locked basin.

A different sequence from the same basin was interpreted as consisting of tidal deposits. Evidence of tidal deposition includes bimodal crossbedding, dominant toward the inferred landward direction. The absence of herringbone crossbedding was specifically noted. Flaser bedding and crossbeds with mud-draped foresets, indicative of fluctuating (tidal) currents, are also present. Oscillation ripples and possible marine trace fossils complete the evidence of marine conditions. The local presence of associated black (anoxic) mudstones supports the viewpoint that this sequence was formed in a subaqueous environment. Associated coarsening-upward sequences of mudstone to rippled sandstone to crossbedded or massive sandstone were interpreted briefly as bay and bay-mouth deposits.

Facies: Patterns and Controls

As shown in the examples above, thick monomineralic deposits of halite may accumulate in marginal-marine settings in contemporaneously subsiding basins. Local lateral facies changes include basin-center potassium and magnesium end-member evaporite minerals that grade into basin-margin sulfate minerals—principally anhydrite/gypsum. An ideal lateral fractionation series of evaporite minerals on a broader regional scale has been difficult to demonstrate convincingly, however, and many geologists have resorted to ambiguous assumptions of previously depleted brines.

Although most of the evidence presented suggests that the Triassic rift-basin halites are marine-influenced, Hardie (1984) has suggested that thick monomineralic halite deposits may also be deposited in nonmarine settings. Such deposits are usually associated with coarse, nonmarine basin-margin clastic deposits, whereas the nature of many Triassic deposits is apparently chemical all the way up to the faulted basin margin. Van Houten (1977, p. 85) has noted that, in the Moroccan salt basins, "a remarkable balance was maintained between rapid precipitation of salt and basin subsidence, uncontaminated by significant influx of detritus." Therefore, this model of thick nonmarine evaporites is not applicable to most of the Triassic halites.

Unfortunately, the scale of facies changes in the Triassic basins across western Europe and North Africa is so great and the patterns are so discontinuous that they cannot be physically traced back to an ultimate marine source of the brines. It appears that the brines were transported for hundreds of kilometers, depositing end-member halites and potash minerals in several intermediate locations. The resulting facies patterns are rather complex. A local, more rapidly subsiding (but not necessarily deep) basin was a prerequisite for the deposition—or at least for the preservation—of chlorides.

Small-scale vertical patterns record episodic deposition and dissolution within these thick salt deposits, rather than lateral facies migrations. Clayey partings may represent "lag" concentrations of siliciclastic material left behind during the dissolution of significant volumes of clay-bearing halite. Vertical facies changes of mineral phases have rarely been reported.

On a broader scale, the thick evaporite deposits commonly grade up out of clastic deposits, thus recording a combination of decreased input from reduced-relief source areas, continued basin subsidence, and (probably) increased marine influence. Only in South Wales has it been suggested that significant topographic relief and nonmarine environments were present at the edges of a basin in which mudstones and evaporites were being deposited (Tucker 1977, 1978).

Summary

The thick evaporitic deposits of the Triassic rift basins in Europe and North Africa are probably marine-influenced sediments, deposited in low-relief settings. Their facies patterns and mineral assemblages are not easily explained by theoretical marine evaporation sequences, however, and they apparently required a specific and delicate balance of tectonic, eustatic, and climatic conditions for their formation. Nonetheless, such conditions were widespread in Europe and North Africa during Late Triassic time. These conditions were not met in the contemporaneous onshore basins of North America, where significantly smaller quantities and significantly different types of evaporites were deposited in significantly different environments.

Data Collection and Interpretation Techniques Illustrated by Studies of Rift-Basin Marine-influenced Deposits

Evaporite deposits often prove difficult to study, both chemically and sedimentologically, because of the ease with which they recrystallize and change mineral phases at low temperatures. These changes destroy the primary sedimentary textures and primary mineral assemblages. Conventional chemical and mineralogical studies are now being augmented by the study of the remnant clastic textures sometimes found in evaporites, however, and environmental interpretations are increasingly being based on such evidence. These latest techniques have not yet been widely applied to the Triassic rift-basin evaporites.

Chemical studies that have been applied to these deposits are few, and the results are often ambiguous. Some geochemical studies have been useful, however. Clemmensen, Holser, and Winter (1984) published a sulfur and carbon isotope study that suggested a nonmarine origin for many of the sulfate deposits found in sabkha deposits in Greenland. Several authors have made use of bromine analyses to distinguish marine from nonmarine halites, but few such analyses are performed regularly. Hardie (1984) has recently suggested that the bromine content of halite is not as definitive a marine versus nonmarine indicator as was once thought, since the ranges of values for the two environments overlap. Exceptionally low bromine content in a halite, however, may be indicative of either nonmarine or recycled marine origin, whereas the consistent bromine values through a large stratigraphic range in the English Triassic halites are probably good evidence for halite deposition from a uniform, marine-sourced brine.

Mineralogically, the Triassic evaporites are relatively simple. The suite of end-member potash minerals present in these deposits in Morocco is not complex compared with other potash-bearing formations. The mineralogical simplicity of most of the halites is, perversely, a source of puzzlement in itself. In the rift basins, the reasons for the anomalously low volume of minerals that precipitated out from evaporating seawater before halite (sulfates and carbonates) are not well understood. Still, no one has yet published an estimate of mineral volumes or mineral balance calculations for these deposits, so it is difficult to ascertain the degree of disparity between the predicted and observed mineral trends. The absence of such distinctive minerals as the sodium-bearing evaporites (glauberite and thenardite), on the other hand, is evidence for relatively elementary brines with simple fractionation and source histories.

Instead of simple lateral fractionation of minerals from brine source to distal basin, the regional mineral facies patterns seem to have been controlled by the more local conditions of water depth, basin subsidence rate, and local influxes of fresh water and clastic sediments. Thick salts were deposited in the more rapidly subsiding grabens, whereas sulfates precipitated in the thinner, muddy sediments between basins. If halite was also deposited in areas between the basins, it was rarely preserved. Carbonates of any kind are notably rare in most cases. Thus, the facies patterns conform poorly to any theoretical sequence. The best "type" facies patterns have been reported from the poorly known Aquitaine basin.

The recently studied clastic structures present in some evaporite deposits (Hardie, Lowenstein, and Spencer 1983; Warren and Kendall 1985) have not yet been recognized in most of the Triassic rift basins, nor have studies of their fluid inclusions been reported on. Detailed studies of textures and fabrics of preserved primary crystals have, however, been used to good advantage by Arthurton (1973). Such studies concentrate on crystal form, crystal orientation, and patterns of fluid inclusions, which reflect growth patterns within the crystals. They also include the associated siliciclastic, muddy sediments and such secondary features as crystal truncation planes caused by dissolution .events. This type of study of sedimentary and crystal structures is an important source of evidence for drawing valid conclusions as to water depth, but few such studies have been published for the Triassic deposits.

Studies of associated sediments are crucial to the interpretation of evaporite deposits. The nature of their sedimentary structures and the trends of these sediments as they grade into the evaporites are critical to environmental interpretations of the usually unfossiliferous and recrystallized evaporites. Additionally, nonsedimentary evidence such as pillow basalts should not be overlooked.

11 / Eolian Deposits

One reason why European geologists did not generally accept Russell's early theory of a warm, humid climate for the origin of redbeds was that they had red Triassic sandstones of undoubtedly eolian origin. Although Reinemund (1955) and Prouty (1931) briefly mentioned isolated occurrences of what they considered to be Triassic ventifacts in North Carolina, true eolian deposits are rare in the North American basins. They have only recently been described from the Fundy Group of maritime Canada (figure 11-1) (Hubert and Mertz 1980). Eolian deposits also occur in England and in the Central East Greenland basin.

As they are currently understood, eolian deposits in the Triassic rift basins are localized, occurring primarily within the zone of trade winds north of the Triassic equator (Hubert and Mertz 1984). They do not all occur at the same stratigraphic interval, however (see table 11-1), indicating that they were not the product of an isolated climatic event.

The Triassic–Jurassic eolian environments were localized dune fields within rift valleys. Their deposits are usually interbedded with deposits from other environments, and apparently they were small- to medium-scale dunes rather than the large draas of extensive sand seas.

Of all the environments considered here, eolian deposits are the least affected by tectonic regime and topography. In addition, eolian deposition is not affected by local or oceanic base levels. On the other hand, the eolian environments are the most dependent on climatic conditions for their formation.

There are numerous types of sand dunes and, therefore, several types of eolian stratification (McKee, ed. 1979). The interdune deposits can exhibit various characteristics, depending on the proximity of the water table to the

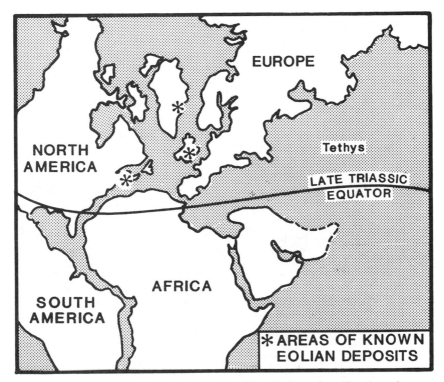

Figure 11-1. Known circum–North Atlantic, Triassic–Jurassic eolian deposits are restricted to the northern rift basins of Greenland, maritime Canada, and England, although there are clues that ephemeral eolianites were present in other basins.

Table 11-1. **Age distribution of known eolian deposits.**

Location	Age	Reference
Nova Scotia	Late Triassic and Early Jurassic	Hubert and Mertz (1984)
Nova Scotia	Early Late Triassic	Nadon and Middleton (1985)
Scotland	Late Triassic	Craig (1965)
Greenland	Middle Triassic	Clemmensen (1978a)
England	Early to early Middle Triassic	Thompson (1969)

Figure 11-2. Close-up of foreset bedding: thin, dark, finer-grained grain-fall laminae interbedded with thicker, coarser, grain-flow laminae, Fundy basin, Canada (coin for scale). (Photo courtesy of J. F. Hubert, from Hubert and Mertz 1984, Figure 4D. Reprinted with the permission of the *Journal of Sedimentary Petrology*)

surface, the degree of aridity of the climate, and the chemistry of the groundwater. The few eolian deposits found in the rift basins exemplify only a small part of the total variability possible within eolian sedimentary deposits.

Eolian Deposits of the Fundy Basin

Perhaps the best sedimentological study of eolian deposits in a Triassic-Jurassic rift basin to date is the one performed by Hubert and Mertz (1980, 1984). Early interpretations for eolian sandstones had been based on the generalized evidence of large-scale crossbeds composed of well-sorted, usually rounded sand. Hubert and Mertz were the first to recognize eolian deposits in the Fundy basin, and were able to apply a detailed knowledge of sedimentary structures to their interpretations, based on studies of modern eolian environments. The sedimentary structures and features they used to interpret the deposits include the following:

1. Subround to round, moderately well- to very well-sorted sand grains. Typically, the sand grains from the eolian deposits were more abraided and finer-grained than those from associated fluvial deposits.
2. Grain-fall and grain-flow laminae as crossbed foresets (figure 11-2). Hubert and Mertz recognized the former as being somewhat finer-grained and better-sorted, whereas the latter often wedged out downward within climbing ripple structures.
3. Climbing-ripple translatent strata within crossbed foresets. These structures were produced as winds blew obliquely across the slipface of the dune. They are more common near the base of the slipfaces, where they often formed slipface toe sets that became tangential to the lower bounding surface. Most of these ripples have a relatively high ripple index (ratio of ripple wavelength to ripple height). Hubert and Mertz (1980) reported an average index of 24, corresponding to modern eolian ripples.
4. Large crossbed sets with high-angle foresets (figure 11-3). The average crossbed thickness in the Fundy basin is 1.2 meters, but thicknesses range up to 3.5 meters. Foreset/slipface angles approach the angle of repose of sand, especially if postdepositional compaction is accounted for.
5. Three orders of bounding surfaces (figure 11-3). The first-order surfaces truncate all sedimentary structures and extend for a minimum of hundreds of meters with only tens of centimeters of relief. Hubert and Mertz attributed these surfaces—which are marked by pebble lags, and often by polished and pitted ventifacts—to flash floods that sculptured the desert floor. Second-order surfaces generally are inclined and planar, dip downwind, and record shifts in the migration direction of individual dunes. Third-order surfaces occur as interruptions within the crossbed sets and are interpreted as reactivation surfaces caused by local wind shifts.
6. Rare associated interdune deposits.

By sketching outcrops and noting the positions and orientations of sedimentary structures, Hubert and Mertz (1984) were able to document several instances of "sweeping arcuate patterns of crossbed azimuths within single sets." This gradual rotation of foreset azimuth along the outcrop indicates

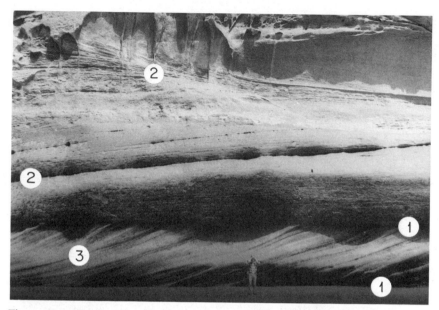

Figure 11-3. Large-scale bedding in eolian deposits, Fundy basin, Canada (geologist for scale). Numbers indicate the heirarchy of the bedding planes: (1) erosion surfaces created by flash floods and subsequent deflation; (2) surfaces bounding crossbed sets created by superimposed slip faces; (3) slip-face crossbedding. (Photo courtesy of J. F. Hubert, from Hubert and Mertz 1984, Figure 4A. Reprinted with the permission of the *Journal of Sedimentary Petrology*)

curved dune crestlines (figure 11-4); thus Hubert and Mertz interpreted the dunes to have been barchans or, where barchans coalesced, barchanoid transverse ridges.

The 180- to 240-degree variation within local crossbed azimuths is consistent with this dune morphology. Hubert and Mertz suggested that a secondary mode within the cumulative crossbedding azimuths, oblique to the main mode, represents a seasonal, temporary change in wind patterns. In addition, they used the observed grain-flow crossbed thicknesses (2 to 3 centimeters) to calculate the original height of the barchans as about 2 to 7 meters, using an empirical ratio published by Kocurek and Dott (1981).

The resulting local paleogeography reconstructed by Hubert and Mertz is one in which dunes formed on the alluvial-fan and fluvial deposits at the edge of the basin. The underlying deposits were the source of the eolian sands. During episodes of high fluvial discharge and progradation of the alluvial fan–fluvial sequence (probably associated with renewed tectonism on the border faults), the eolian deposits were either eroded and reworked or buried and preserved.

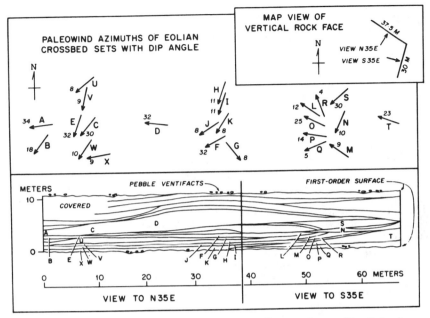

Figure 11-4. Sketch of vertical outcrop faces of eolian sandstone, Fundy basin, Canada. Lettered beds are crossbed sets that correspond to the azimuth vectors plotted above the sketch. Note that the azimuths rotate systematically along each crossbed set, implying an arcuate slipface and dune crest, and thus a barchan type of dune. (From Hubert and Mertz 1984, Figure 11. Reprinted with the permission of the *Journal of Sedimentary Petrology*)

Nadon and Middleton (1985) used an assemblage of general criteria similar to those used by Hubert and Mertz to identify eolian sediments in another part of the Fundy basin, but they concluded that the evidence was insufficient for determining dune morphology.

Eolian Deposits in England and Morocco

Thompson (1969), in another excellent study of eolian deposits, drew an analogy between the internal stratification of some Triassic sandstones in the Cheshire basin of England and the stratification in the modern gypsum-sand dunes of the White Sands of New Mexico, which had recently been described (McKee 1966). Much of this comparison was based on the characteristics of large-scale cross-stratification. Thompson postulated complex "dome-shaped" dunes, based on detailed sketches of his outcrops that showed extensive low-angle crossbedding and interbedded, smaller, high-angle crossbed sets (figure 11-5).

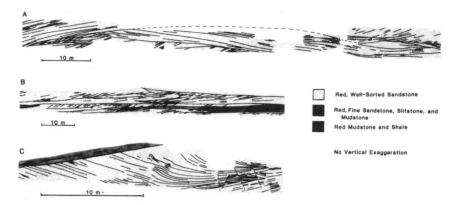

Figure 11-5. Examples of the extensive two-dimensional exposures of dune structures traced from photographs and used by Thompson to infer "dome-shaped" dunes in the Cheshire basin, England: (A) probable shape of convex-up dome-shaped erosion surface; (B) "sub-dune" complex, with interdune deposits; (C) large, sweeping, asymptotic foresets (middle and left) of dome-shaped dune, encroaching over smaller crossbeds of second dune (lower right). Note the contorted foresets. (From Thompson 1969, Figures 4 and 5. Reproduced with the permission of *Sedimentary Geology*)

Thompson's detailed study of the arrangement of well-exposed sedimentary structures in eight fossil dunes in the Cheshire basin showed that the core of the dunes, and (locally) the downwind areas contain large-scale, high-angle, slipface crossbedding typical of transverse dunes. Superimposed on these initial core structures, however, are the lower-angle eolian bed forms, smaller crossbeds, and erosional blowout structures indicative of very strong, irregular winds.

Thompson proposed that transverse dunes were formed from sand derived from nearby dry riverbeds during periods of moderate winds, and that during subsequent gale-force wind conditions, the transverse dunes were degraded to dome-shaped dunes. Degradation consisted of modification of the upper parts of the dunes by erosion and superposition of smaller, irregularly oriented crossbeds, and of elongation of lee slopes into lower-angle, concave-upward, grain-fall slopes. Simultaneously, the plan form of the dunes (as indicated by systematic changes of crossbed vectors along the outcrop) was modified to the convex-downwind profile of dome-shaped dunes. Subsequent moderate wind conditions allowed transverse slipfaces to re-form locally on the downwind edges of the dunes.

Thompson (1970*b*) also noted an inverse correlation between the presence of mica flakes and the presence of "millet-seed grains" (well-rounded spherical grains) in these sandstones. He found that micas—and, to a lesser extent, feldspars—were more common in associated fluvial deposits, from which the otherwise similar petrography showed the eolian sands to have been derived. Because of abrasion and deflation(?), however, these grain types were rare in the eolian environments that were capable of producing the well-sorted, millet-seed sandstones.

Henson (1970) wrote briefly of eolian dune deposits in South Devon, England. These were recognized on the basis of sandstones with large ("dune-scale") crossbedding and little mica. Because of the sporadic occurrence of these deposits, the eolian environments were interpreted as having been isolated dunes and restricted dune fields that were derived from the fan deposits on which they formed—similar to the dunes in the Fundy and Cheshire basins.

Henson (1970, p. 175) also noted a "deflation layer of ventifacts up to 7 centimeters thick." The ventifacts were not described, but an interesting point was made that the formation immediately below this layer is dominantly sandstone, the implication being that the ventifact layer represents the deflation of a considerable thickness of sandy deposits in order to concentrate a 7-centimeter-thick layer of pebbles.

No eolian deposits have yet been recognized from the rift basins of Morocco, although they are thought to have been located at about the same paleolatitude as the Fundy basin. The ephemeral existence of eolianites in the Kerrouchen basin is strongly suggested, however, by the local presence of exceptionally well-rounded sand grains in some of the fluvial and flood-plain deposits (figure 11-6). Eolian rounding occurred despite the proximity of the deposits to their granitic/metamorphic source terrane (Lorenz 1976).

Interdune Deposits

Whereas Hubert and Mertz noted a definite paucity of interdune sediments in their area of the Fundy basin, Nadon and Middleton (1985) found a number of such deposits interbedded within the dune sandstones in a nearby area of the same basin. Interdune deposits have also been described in England by Thompson (1969) and in Greenland by Clemmensen (1978a). These types of deposits have been recognized on the basis of two characteristics:

1. Presumed adhesion ripples. These irregular laminations are thought to have formed as wind-blown sand was trapped on the moist surfaces of the low-lying interdune areas. As noted earlier, many such features may also be attributable to minor salt-disturbance features. In Canada (Nadon and Middleton 1985), beds of horizontally bedded to massive sandstone are thought to have been deposited in these zones.

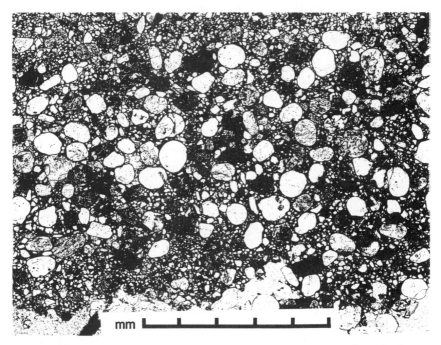

Figure 11-6. Well-rounded sand grains mixed with much smaller silt grains in a polymodal grain-size fabric, suggesting eolian influence in the environment in which this thin, relatively structureless bed was deposited, Kerrouchen basin, Morocco. (Photo by the author)

2.   Red and brown siltstones and shales, often interbedded with each other and often bearing desiccation cracks. Such deposits are usually taken to represent interdune ponds and/or sabkhas, as the fine-grained sediments would otherwise have been blown away. Clemmensen (1978a) has reported on one such suite of deposits that are gypsiferous, indicating intense evaporation from these areas. These sediments were probably the source of the gypsum sand that formed adjacent dune deposits.

   Although high ripple-index numbers are one of the most characteristic features of eolian deposits, few such measurements have been reported. In theory, interdune deposits should contain abundant high-index ripples, and the deposits of damper interdune areas should be burrowed or rooted to some degree. Thompson (1969) did note briefly that the bottom-set sandstones of the dune facies in the Cheshire basin grade laterally into ripple- and flat-bedded mica-rich sandstones and mudstones. He interpreted these as having been "water-laid" in pools and lakes.

Facies: Patterns and Controls

Eolian deposits in the Triassic–Jurassic rift basins interfinger intimately with deposits from several other environments. In the Fundy basin, the dune facies are interbedded with deposits of the alluvial fans and braided rivers on which they formed and from which they derived their sediment. Interfingering of the two types of deposits is suggested to have been caused by alternate episodes of tectonic stability (during which the dunes were allowed to grow), and basin-margin faulting with fan progradation (Nadon and Middleton 1985).

The dunes in the Fundy basin may have been formed preferentially on one side of the basin, possibly as a result of prevailing winds that concentrated wind-blown sand toward the northern bounding highlands. Thus, climate was a large factor in controlling both the existence and the location of dune sediments. The wind and its deposits do not respond to the same controls as do water-born deposits, however, and therefore eolian deposits may be mixed with numerous seemingly unrelated, yet contemporaneous deposits.

In eastern Greenland, Clemmensen (1978a) has shown that the dune facies alternate with evaporitic mudstones that were interpreted as being of nonmarine (playa) origin. Clemmensen reconstructed the local paleoenvironment as one in which the dunes were located as a facies intermediate between the basin-margin alluvial fans and the basin-center playas. Alternate episodes of dune and playa deposition present in measured vertical sections were interpreted as recording variable climatic conditions. Playas expanded during periods of increased precipitation, while the dune fields expanded during drier episodes.

Summary

Triassic–Jurassic eolian deposits are found at a number of stratigraphic intervals, but their presence is limited to the northern basins. This limitation is probably due to the distribution of global wind-circulation patterns and to the necessity of strong, low-variance wind conditions for the initiation of eolian environments. Nevertheless, the rift-basin eolian deposits never developed into large sand seas that dominated the basins. Rather, the dunes were smaller barchan and "dome-shaped" features, now interbedded with noneolian deposits that formed contemporaneously with the dunes but were responsive to different controls on sedimentation.

Interruptions in eolian deposition are recorded by deflation surfaces and lags, by flood-scour surfaces, and, in Morocco, by the complete reworking of the eolian deposits. Less important hiatuses are recorded within the dune deposits as multiple orders of bounding surfaces.

Although the Triassic–Jurassic dunes were small, their sedimentary records exhibit many of the characteristics of modern eolian sediments. These include, among other features, ventifacts, well-rounded sand grains, interdune deposits, and large grain-fall and grain-flow slipface crossbedding; thus, there can be little doubt as to the eolian nature of their depositional environment.

Data Collection and Interpretation Techniques Illustrated by Studies of Rift-Basin Eolian Deposits

The known eolian deposits that formed in the Triassic rift basins are relatively small and uncomplicated, but they contain excellent examples of smaller-scale eolian sedimentary structures. Unfortunately, this scale of structure occurs in all types of eolian dunes, and it is the larger outcrop-scale structures that must be used in the reconstruction of dune morphologies. Here, the geologist is hampered by the difference between the most common vertical two-dimensional outcrops of ancient rocks and the usual horizontal two-dimensional exposures of modern environments. Vertical trenching in modern environments cannot approximate the vertical scale of most outcrops—although, for isolated bed forms, shallow trenches at different points on the dune provide excellent data on the internal geometry within the dune (Krystinik 1986). Any reconstruction of dune morphology must also take into account the low preservation potential of the upper parts of dunes.

The older, generalized criteria for the recognition of eolian sandstones are still valid, but probably with a caveat that the roundness of the grains and their degree of sorting are usually relative, rather than absolute, values. These characteristics in suspected eolian sediments should be compared with the grain textures in associated fluvial sediments. In deposits that have not undergone extensive diagenesis, the relative degree of abrasion of the grain surfaces (Hubert and Mertz 1984) and the relatively low percentage of grains that are susceptible to mechanical attrition (Thompson 1969) may also provide evidence of eolian origin.

The subtle nose-to-the-outcrop-scale sedimentary structures have provided definitive evidence for the presence of eolian deposits. These include crossbed foresets with grain-fall and grain-flow laminae, high-index translatent ripples within the foresets, and perhaps the ill-defined adhesion ripples. Deflation lags and water-table deflation surfaces are also diagnostic of the eolian environments of the Triassic–Jurassic rift basins. Interdune deposits are usually finer-grained, rippled, horizontally bedded, and/or structureless. They are also commonly evaporitic.

Vertical or horizontal facies patterns are not uniquely indicative of eolian depositional conditions. Dune sediments can occur within a number of associated lithofacies because their distribution is not controlled directly by the same factors that govern deposition in associated environments.

Where conditions permit, one of the more promising techniques for determining dune geometry is the sketching of outcrops, with notations as to the distribution of sedimentary structures, their arrangement into units within the bedding, and the orientation and distribution of directional features. Photographs may be of secondary importance to a good sketch, inasmuch as with pictures the data often go directly from the outcrop to the photograph without necessarily passing through the mind of the geologist. (This is an observation from personal experience.) Moreover, directional features cannot be measured back in the office from pictures. Photographs are capable of recording data that may go overlooked in the process of sketching, however, and they record information without the filter of scientific bias.

Finally, the presence of deflation surfaces is an important characteristic of these easily eroded sediments. These may be marked by pebble lags wherein the pebbles are characteristically spaced and often pitted, polished, and faceted. They are often marked by the extensive low-relief surfaces that truncate sedimentary structures, indicative of deflation down to the level of the water table and/or of flash flooding.

12 / Overview of Triassic–Jurassic Rift-Basin Sedimentology

Even though an attempt has been made to present all of the evidence on which different authors have based their conclusions, some of the preceding chapters are short because the relevant published data are minimal. In fact, the ratio of evidence to conclusions is often disconcertingly low. Moreover, numerous Triassic–Jurassic basins contain strata that have not yet been the subject of study with up-to-date techniques of sedimentological investigation. Because techniques and theories are always changing, there will always be a limited number of entirely up-to-date studies. Several distinctive and repetitive types of lithologic suites are represented in the rift basins, however— from the conglomeratic sediments (interpreted as subaerial alluvial-fan deposits) to thick beds of halite that were probably deposited in periodically inundated salinas, with a wide range of variability within each category.

This chapter first examines some of the regional factors that controlled the patterns of Triassic rift-basin sedimentation; that is, it examines why every rift basin did not produce fault-margin alluvial fans that graded distally into fluvial and then into playa environments. This chapter also reviews features of each environment that are both preserved and diagnostic. Features of modern analogues that seem to have little use in the reconstruction of the paleogeography and paleoenvironment are noted.

The usefulness of the techniques of data collection and interpretation that have been illustrated are assessed here. And finally, the general scientific methods applicable to the historical science of sedimentology are examined.

Regional Controls on Rift-Basin Sedimentology

The most significant factor controlling the type and configuration of deposits that filled the Triassic–Jurassic rift basins was plate tectonics. Rates of plate motion probably governed eustatic sea level and the ability of seas to flood areas of the continents. The interaction and breakup of plates controlled vertical relief on the continents, both regionally (where broad cratonic lowering facilitated inundations of sea waters) and locally (where faulting created high-relief source areas and adjacent rift basins). Moreover, if current models of rift processes are correct, regional paleoslopes on the continental margins varied as a function of the stage of continental breakup. Initial slopes were probably toward the main rifting axis, but postrift paleoslopes may have been reversed. After rifting, a flexural bulge with an amplitude of some 50 meters (Watts and Ryan 1976) slowly migrated across the continental margin (figure 12-1), possibly affecting paleoslopes within the basins and/or postdepositional erosion.

The arrangement of plates with respect to each other and with respect to global climatic belts caused variations in local climatic conditions: inland areas probably had the driest of the generally arid climates that prevailed in the Triassic, while rift basins that were located along the belts of consistent winds were probably more susceptible to eolian deposition. Finally, tectonics controlled the syndepositional volcanism that occurred in many of the basins, the effects of which on sedimentation are poorly understood. These controls on sedimentation are examined below, under the headings of climate and paleogeography, eustatic sea level, and tectonics.

Climate and Paleogeography

The overall climate of any paleoenvironment of deposition is hard to reconstruct. Robinson (1973) and Habicht (1979) have summarized the general consensus that the Triassic era was one of predominantly arid climates. The evidence consists primarily of lithologic clues, such as the volumetric importance of redbeds and evaporites and the presence of eolian deposits in Triassic deposits worldwide. It might therefore be considered circular reasoning to infer that one of the effects of the arid Triassic climate on sedimentation was to produce redbeds, evaporites, and eolian deposits, but this is more a case of mutually supporting observations than of circular reasoning. The recent recognition of abundant caliche paleosols in many of the Triassic deposits has independently strengthened the case for an arid to semiarid climate.

Other indicators do not immediately fit into an arid climatic reconstruction. The Triassic coals of the central Atlantic states of the United States, the lacustrine deposits, and the abundant flora and aquatic fauna represented at some stratigraphic intervals of the North American basins require locally abundant water. These may have been geographically controlled—for exam-

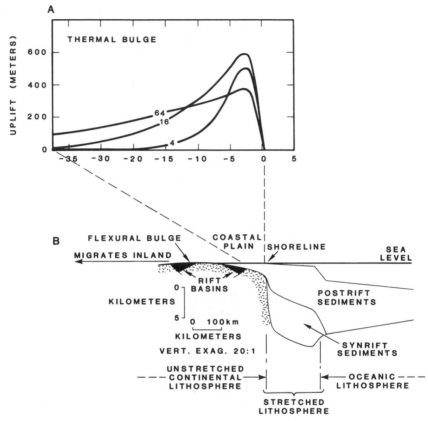

Figure 12-1. During initial stages of rifting, the lithosphere subsides as it is thinned and stretched, and regional paleoslopes in the area of the rift basins are toward the center of subsidence. Once actual separation of the two new continents occurs, the new continental margins are uplifted (A) as a result of thermal effects, reversing the regional paleoslopes. Numbered curves indicate theoretical land surfaces at different ages (in millions of years after separation). As the passive continental margin is loaded with sediments, it is depressed (B) and subsidence overwhelms the thermal bulge. This loading and depression of the continental margin also causes a crustal warp or flexural bulge inland. As the postrift lithosphere cools, its flexural strength (and thus the wavelength of flexure) increases, causing the position of the flexural bulge to migrate inland.

Phases of uplift, and phases of migration of a bulge in the lithosphere across the surficial rocks within the rift basins, could cause significant erosion and structural deformation of the strata, possibly affecting patterns of sedimentation within the basins (if they occur syndepositionally). (After Watts 1981, Figure 24, and Watts 1982, Figure 2. Reproduced with the permission of A. B. Watts and *Nature*, Macmillan Journals Limited)

ple, by local phenomena such as rivers flowing across semiarid to arid plains toward basins of internal drainage—rather than being controlled by major climatic excursions, although some of the lacustrine cycles may have been directly controlled by climate. Manspeizer (1982) has described rapid local climatic variations caused by the high-relief topography adjacent to modern rift-basin settings and has inferred analogies to the Triassic-Jurassic basins.

The reconstructed geochemistry of lakes in these basins of internal drainage suggests that they were strongly affected by an evaporitic climate. Rather than filling to a spillover point, lake waters were kept at a low level by evaporation, and the concentrated dissolved components of the source area precipitated as exotic authigenic zeolites and evaporites.

The coals of the mid-Atlantic states of the United States are among the oldest (Middle Triassic) of the North American rift-basin deposits, which might suggest that the overall climate became more arid through Triassic time, but such a trend is not reflected in contemporaneous European deposits. Clemmensen (1980) has in fact suggested that younger coal deposits of Triassic–Jurassic transition age in Greenland indicate that the climate became more humid at the end of Triassic time. As noted earlier, however, coals are probably not as definitive indicators of climate as they are often assumed to be.

The occurrences of eolian dune fields also seem to have been geographic perturbations on a generally arid climate. Eolian deposits would otherwise have been more widespread and would not have been as sporadically distributed through Triassic time (see table 11-1). The presence of steady, strong prevailing wind patterns in low-latitude arid belts may have been instrumental in creating local eolian deposits within the overall arid climate (Waugh 1973).

The location of most of the basins toward the interior of the supercontinent of Laurasia would have put most of them in an unfavorable location for receiving whatever precipitation was available (figure 12-2). Moisture-laden air currents originating over distant seas would have had to travel far inland and would probably have lost much of their moisture before reaching these rift basins (Robinson 1973). Little enough humidity and precipitation were available to begin with, however, as evidenced by evaporitic deposits in rift basins that were close enough to the sea (within a few hundreds of kilometers?) to receive nfluxes of marine brines and yet maintained high evaporation-to-precipitation ratios.

Minor fluctuations evidently took place within this overall arid climate. The lacustrine cycles of transgression and regression (or desiccation) are probably results of such fluctuations, although the absolute origin of the lacustrine conditions may not have been climatic. Bluck (1965) and LeTourneau (1985a,

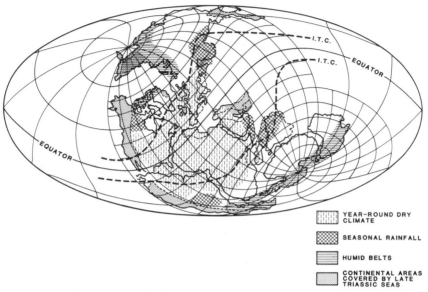

YEAR-ROUND DRY CLIMATE

SEASONAL RAINFALL

HUMID BELTS

CONTINENTAL AREAS COVERED BY LATE TRIASSIC SEAS

Figure 12-2. Reconstruction of the course of the intertropical convergence zone (I.T.C.) for July and January, and major climatic regions for combined Late Triassic plate configurations and global air circulation patterns. The I.T.C. zone is an area of ascending air that can produce rain belts if the associated trade winds have passed over oceanic regions. Its position shifts northward in summer and southward in winter. (From Robinson 1973, Figure 10. Reproduced with the permission of Academic Press)

1985*c*) have suggested that responses to these minor climatic fluctuations can be found in sediments other than lacustrine ones. They have reported alternating assemblages of lithologies in alluvial-fan deposits: debris flows and ephemeral braided-stream deposits are considered to have formed under relatively arid conditions, whereas the deposits of less intermittent streams formed during less arid climatic intervals. Distally, such variations produced alternating lacustrine and playa environments.

Long-range, tectonically caused climatic changes may also have occurred through Triassic–Early Jurassic time. Although most of the basins were located within 20 degrees of the Late Triassic equator, motion of the plates carried them across about 15 to 20 degrees of latitude and through several possible climatic zones as sedimentation progressed (Manspeizer 1982). Manspeizer has suggested that this motion and the attendant climatic shifts are documented in the stratigraphic record within the rift basins.

Thus, climatic conditions exerted a strong influence on Triassic rift-basin

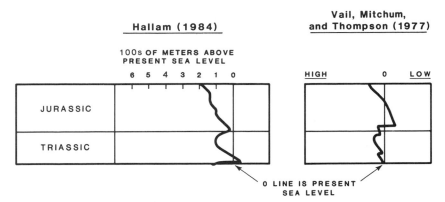

Figure 12-3. Different interpretations of changes of eustatic sea level through Triassic and Jurassic time: (A) after Hallam 1984; (B) after Vail, Mitchum, and Thompson 1977. The Triassic-Jurassic fluctuations are minimal compared to the entire range of inferred variation through the Phanerozoic. Note that the timing of sea-level rises does not correspond well with the Triassic transgressions in Europe and North Africa. (After Hallam 1984, Figure 5. Reproduced with the permission of the American Association of Petroleum Geologists and the *Annual Review of Earth and Planetary Sciences*, Volume 12, copyright 1984, by Annual Reviews, Inc.)

sedimentation. Saline, alkaline lakes developed where the same conditions might have produced freshwater lakes in a different climate, and playas formed where saline permanent lakes might otherwise have been the norm.

Eustatic Sea Level

Reconstructions indicate that eustatic sea level rose either smoothly (Hallam 1984) or erratically (Vail, Mitchum, and Thompson 1977) during Triassic time, and then fell more rapidly either in Late Triassic or Early Jurassic time (figure 12-3). The maximum level of transgression is estimated to have been only on the order of 200 meters higher than present sea level (Hallam 1984), however, and sea level was probably generally lower during Triassic time than at any other time during the Phanerozoic except the present.

Moreover, the Triassic sea level may have been falling at the time of maximum marine influence (latest Triassic and earliest Jurassic) in the rift basins of Europe and North Africa, whereas previously, nonmarine to paralic conditions were prevalent in most of the rift basins contemporaneously with the height of the Middle to Late Triassic transgression. Thus, eustatic sea level was less influential on Triassic rift-basin sedimentation than was regional tectonism. Regional subsidence in Europe and North Africa during the Late Triassic apparently occurred on a scale and at a rate that equaled or exceeded the contemporaneous fall in eustatic sea level, although a close proximity of the two must have been maintained for some length of time (Van Houten 1977).

Tectonics

Besides governing the distribution of continents and the regional subsidence of Europe and North Africa (in an as-yet undefined way), tectonism played a very strong local role in the sedimentation of the Triassic–Jurassic rift basins. The principal effect of tectonism was to create significant local relief between the subsiding rift basin and the adjacent source areas. Sedimentation on the basin-margin alluvial fans responded directly to this relief. High relief produced rapid erosion and coarse deposits on the fans. As erosion wore down the relief of the source area, the deposits on the fans became correspondingly finer-grained. Renewed tectonism reactivated old faults (or created new ones), restored high-relief conditions, and caused a return to coarse-grained deposition. Such sequences of tectonic activity can be recognized in basin-margin unconformities, and in repeated gross-scale (hundreds of meters), fining-upward trends in alluvial-fan deposits—especially if the trends can be correlated within the basin and thus distinguished from autocyclic processes.

Exceptions to this seem to have occurred in some of the rift basins of northern Morocco, where very little conglomerate or clastic sediment was deposited despite large offsets on basin-margin faults. In these basins, two modifications to the general model must have taken place: uplift of the area adjacent to the graben must have been minimal, so that all the throw on the fault was accommodated by subsidence of the floor of the basin; and sedimentation of evaporites and mudstones within the basin must have kept pace with subsidence, so that relief between the basin floor and the adjacent areas was always minimal, and little coarse, clastic sediment was produced.

On a larger scale, the tectonic regimes of England, Iberia, and North Africa differed from those of North America in several ways. The rate of tectonism seems to have decreased during the Triassic in basins on the eastern side of the Atlantic, such that the initially coarse-grained graben fills (some of which may be as old as Permian) gave way to the finer-grained deposits of more gradual and regional subsidence. On the western side of the Atlantic, the North American rift basins did not form until later in the Triassic, but active faulting and coarse-grained, alluvial-fan deposition often extended into Early or perhaps even Middle Jurassic time. Intermittent tectonically inactive stages are recorded in North America, as a series of fining-upward, basin-fill sequences.

Many of the fine-grained lacustrine sequences in the basins may indicate local or regional tectonic rearrangements of paleoslopes, rather than tectonic stability. Alternatively, lacustrine conditions may have originated as a result of irregular tectonic subsidence of a nonhomogeneous basement.

This nonhomogeneity is the result of the structural fabric imprinted on the basement by the previous (Variscan) orogeny. The structural fabric also controlled the location and orientation of the basins. Triassic basin-margin faults formed preferentially along inherited structures despite the orientation of Triassic stresses oblique to most such structures (King 1961; Swanson 1986).

Some components of the Triassic rift-basin tectonic regimes have not yet been fully investigated. These include what appear to be regional subsidence trends in Europe and North Africa and the subtle effects of regional rifting processes on the North American basins.

Data Collection and Analysis

The collection of data and their analysis are integrated activities and are difficult to separate as topics for discussion. Collecting data without having a specific use in mind for those data is less efficient than collecting data with an eye toward their use in solving a particular problem. On the other hand, miscellaneous data often have application to a solution in hindsight, or may even expose unexpected problems.

Many techniques of data collection and analysis are illustrated by sedimentological studies of the Triassic rift basins. Some of the specifics have been covered in earlier chapters; a more general review is given here.

Mapping

Mapping, on a scale commensurate with the variability of whatever is being mapped, provides a solid basis for understanding the geology in question. Numerous types of geological data lend themselves to mapping, including lithology, facies, grain-size data, formation or facies thicknesses, and directional (paleoflow and/or paleoslope) features. All of these have been used successfully in paleogeographic reconstructions and sedimentological interpretations in the rift-basin deposits. Clast-size distribution maps (see, for example, figures 5-10 and 5-13) have been especially illuminating in bringing scattered outcrop data into focus by outlining alluvial-fan positions and shapes. Maps of crossbedding vectors from given stratigraphic horizons (for example, in Hubert et al. 1978) are also invaluable in the reconstruction of paleogeography and in basin analysis.

Isopach maps of the variability in coal and evaporite deposits (see, for example, figures 7-5 and 10-9) have proved useful in the synthesis of subsurface data into paleogeographic concepts. These deposits are usually poorly exposed in outcrop, and isopach trends derived from drillholes provide alternative approaches to sedimentological studies of outcrops. Otherwise, subsurface maps have been little used in rift-basin studies because of the scarcity of drillholes.

Maps of lithologies that do not account for time lines (such as Emerson's 1917 map of the Massachusetts part of the Hartford basin) can be misleading. Lithologies must be mapped in accordance with stratigraphic principles. Lithologic maps of individual 7.5-minute quadrangles (1:24,000 scale) are usually too small-scale to be of use in basin analysis studies. They do add invaluable data to the overall data base, however—especially when enough adjacent maps have been published to make possible the synthesis of a more regional picture.

Cross Sections

Cross sections are essentially maps on a vertical plane, although most of them are more interpretive than plan-view maps are. Cross sections of basins have often been an exercise in synthesis by themselves, requiring the geologist to interpolate between known suites of data on the basis of geologically sound principles of correlation in order to fill in the gaps. Cross sections have been put together from outcrop data, from subsurface data, and sometimes from a combination of both.

Cross sections can illustrate facies trends and facies equivalencies for paleogeographic reconstructions. They can also serve to highlight important thickness variations of both bed- and formation-scale units (for example, the westward thinning of lacustrine deposits in the Hartford basin; see figure 8-8), and thus they may be useful in deciphering the tectonic patterns in the basins. Where a basalt-flow datum is present, well-controlled cross sections provide an excellent portrayal of the lateral extent of specific environments (for example, the limited lateral extent of lacustrine facies at a given stratigraphic level: Demicco and Gierlowski-Kordesch 1986), as well as an idea of small-scale thickness variations for formations. Fence diagrams (interconnected, three-dimensional cross sections) represent another aspect of data portrayal (figure 12-4), but they have been underutilized in most studies of the rift basins. Cross sections down the axis of a basin are much less common than transverse cross sections of most basins, despite their potential usefulness.

Vertical Profiles

Vertical profiles provide much more local (but potentially much more detailed) information than do cross sections or maps. Vertical profiles may

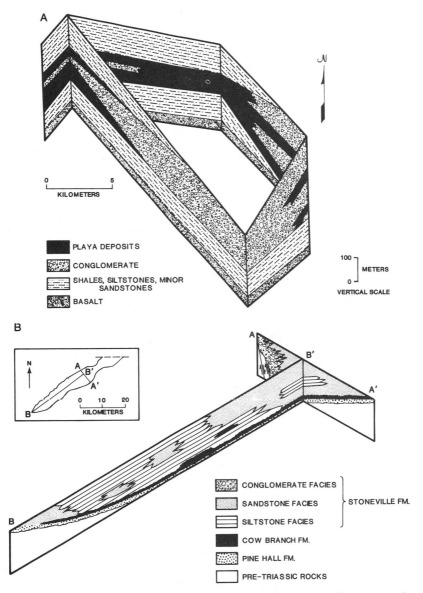

A

0   5
KILOMETERS

■ PLAYA DEPOSITS

CONGLOMERATE

SHALES, SILTSTONES, MINOR
SANDSTONES

BASALT

100 ⊤ METERS
0 ⊥
VERTICAL SCALE

N

B

N
A
B'
A'
B
0   10   20
KILOMETERS

A
B'
A'
B

CONGLOMERATE FACIES ⎫
SANDSTONE FACIES   ⎬ STONEVILLE FM.
SILTSTONE FACIES   ⎭

■ COW BRANCH FM.

PINE HALL FM.

□ PRE-TRIASSIC ROCKS

Figure 12-4. Examples of fence diagrams that have been used to interpret and convey facies relationships in rift-basin deposits. (A) Fence diagram controlled by measured sections, restored to syndepositional configuration, showing alluvial-fan conglomerates interfingering with playa mudstones. (From Mattis 1977, Figure 4. Reproduced with the permission of the *Journal of Sedimentary Petrology*) (B) Schematic fence diagram showing stratigraphic and facies relationships across and along the Dan River basin, North Carolina, showing the present structural configuration. (After Thayer 1970, Figure 4. Reproduced with the permission of *Southeastern Geology*)

be of a scale large enough to comprehend an entire basin-fill sequence or of a scale small enough to depict variability within a single bedding unit. Large-scale profiles can illustrate variability of grain sizes, facies, and lithologies through time, as controlled by climates, tectonics, or perhaps autocyclic processes within the basin. Many alluvial-fan deposits display repetitive, large-scale, fining-upward, grain size trends that have been inferred to record alternate episodes of fault activity/high-relief source areas and of tectonic quiescence/source area erosion to low relief conditions (see, for example, figure 5-11).

At the smaller-scale end of the vertical-profile continuum, trends in vertical grain size and distributions of sedimentary structures through individual beds have provided data for interpretating local depositional environments. Fining-upward grain sizes that correlate with successively smaller-scale sedimentary structures in sandstones can be used to corroborate interpretations of meandering fluvial streams—or even to infer them in the absence of other exposed sedimentary structures. Such sequences have less definitive trends in braided fluvial deposits; consequently, vertical profiles have rarely been applied in interpretations of these sandstones.

Vertical profiles that illustrate fine-scale detail in lithologic variations have been instrumental in the development of lacustrine hypotheses (see, for example, figures 4-6 and 8-13). These profiles have provided the basis for recognizing repetitive assemblages of lithologies that might otherwise have been overlooked.

Assemblages

The recognition of distinctive suites or assemblages of related sedimentary structures and lithologies has occurred on several scales. Conglomerates of debris flows and braided streams are interpreted to be alluvial-fan deposits, whereas suites of black microlaminated mudstones and gray laminated mudstones are interpreted to be lacustrine sediments.

On another scale, studies such as that of LeTourneau and McDonald (1985) have suggested that there may be distinctive subassemblages of related facies within overall alluvial-fan to playa/lacustrine systems. Thus, fan lithologies of perennial streams have been correlated with lacustrine sediments and have been interpreted as being the product of a less arid climate, while a subassemblage of debris-flow-dominated fans and playa deposits have been related to a more arid climatic setting (figure 12-5). Similarly, Thompson (1969) was able to recognize different assemblages of eolian sedimentary structures in the Cheshire basin, and to infer dramatically different wind conditions for the two suites of structures.

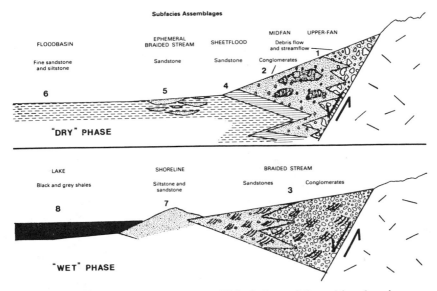

Figure 12-5. Two distinct assemblages of lithofacies and depositional environments that may have been deposited as the result of different climatic regimes. (From LeTourneau and McDonald 1975, Figure 7. Reprinted with the permission of the Connecticut State Geological and Natural History Survey and the U.S. Geological Survey)

Assemblages and arrangements of sedimentary structures of outcrop scale have been neglected in many studies of fluvial deposits. While assemblages have often been noted, the distribution of their components is seldom portrayed in published work. Valuable information in the form of "cross sections" of outcrops can be obtained from sketches, photomosaics, or detailed, closely spaced, correlated, measured sections of single outcrops (see figure 6-1). Ramos, Sopeña, and Perez-Arlucea (1986) and Hubert and Hyde (1984) have presented and used this type of data for fluvial deposits in Spain and eolian deposits in the Fundy basin, but the technique is much more widely applicable. It is basically Miall's (1985) technique of "architectural-element analysis," although it need not be limited to a given set of known sedimentary-structure building blocks.

On yet a larger scale, it has been interesting to compare the assemblages of basin-fill lithologies among the different basins discussed here. Despite the numerous trans-Atlantic similarities, it is evident that the North American rift basins were filled with a somewhat different suite of deposits than those of the eastern margins of the Atlantic. Whereas the North American basins saw repeated episodes of alluvial-fan, lacustrine, and localized playa sedimentation, in the European and North African basins early alluvial-fan and fluvial deposition gave way to the accumulation of graben-filling halites and more regional evaporitic mudstones. Such differences highlight the large-scale variations in tectonic regimes.

Comparisons with Facies Models

Many of the lithofacies of the Triassic–Jurassic rift basins correlate well with existing facies models and with analogous modern depositional environments. These include alluvial-fan, eolian, certain subsets of fluvial, and (probably) playa deposits. Present studies of the marine-influenced, paludal, and braided fluvial deposits are less definitively analogous to facies models of these environments. It is difficult to assess the true correlation of the lacustrine deposits to a facies model, because much of the present lacustrine model (as in Collinson 1978) is in fact based on these same Triassic-Jurassic rift-basin deposits.

In all of these facies models, recognition of the significance of primary sedimentary structures, such as crossbeds and parting lineation, is a prerequisite for reassembly of the specific depositional environment represented by the lithofacies. There remains much room for improvement in making fine-scale distinctions in this area. Ripple marks are commonly noted, but often the type of ripple mark is not. The type of crossbedding in a Triassic deposit has often been noted (for example, trough crossbeds and tabular crossbeds in different parts of alluvial fans: Bluck 1965), but the significance of the differences is still not satisfactorily explained in many modern environments, and therefore little significance has been attached to such observations in the rift-basin strata. Few primary sedimentary structures have been described at all from the cores or from the rare outcrops of the thick halite deposits.

Secondary subaerial modifications to primary sedimentary deposits (such as crumb fabrics and caliche paleosols) are slowly becoming an important part of many facies models, and knowledge of these processes is being applied to Triassic-Jurassic deposits. Some of this activity is due to primary studies of the deposits themselves—much as the lacustrine models are based in part on the Triassic–Jurassic strata.

Methods of study of coal deposits and bedded halites commonly focus on the application of facies models to the associated clastic deposits, rather than to the coals and halites themselves. This type of study is well established in other basins, but it is still in a developmental stage for most of the deposits considered here.

Geochemistry

The geochemistry of the Triassic rift-basin sedimentary rocks has been addressed here only insofar as it has aided in formulating sedimentological interpretations. Most of the geochemical studies noted have been involved with interpretations of evaporites and evaporite-bearing sequences. The bromine content of some Triassic halites and the isotopic composition of the associated evaporites have been analyzed and used as evidence for the marine origin of the brines from which these chemical rocks were precipitated. A detailed mineral-balance study of the evaporite mineralogy has yet to be published, however, and we are left with only the theoretical concepts of concentrated/depleted brines and lateral fractionation.

The playa evaporite minerals in North American basins offer a striking contrast to the more definitively marine-influenced evaporites of Europe and North Africa. The former consist of uncommon minerals such as thenardite, analcime, and glauberite, with only traces of halite, while the latter contain thick, but mineralogically simple sequences of sulfates and bedded halite. The difference between the two suites can be directly related to the different sources of water and conditions of deposition.

The mineralogy of the rift-basin clay deposits has also been addressed in connection with reconstruction of lake-water geochemistry. Detrital clays and other minerals often reacted with hypersaline and alkaline lake waters to produce authigenic and diagenetic clay minerals, and sometimes zeolites (Van Houten 1962, 1964). Concentrated brines can be shown to have affected the mineralogy of the clays in playa environments, whereas the clays of the floodplains have retained much of their primary detrital composition (April 1981).

In a less extreme scale of alkalinity and salinity, Triassic lake-water geochemistries have been reconstructed on the basis of the evidence of minerals and sediments that formed in place, such as oolites and chert. Additionally, the presence or absence of fossil remains of plants and animals has been used to infer whether or not lake waters were compatible with the narrow range of tolerances possessed by most life forms.

Fossils and Trace Fossils

By far the most evidence of Triassic–Jurassic life in the rift-basin strata comes from trace fossils. This is due to poor preservation of organic remains in an oxidizing and generally semiarid climate. Certainly the profusion of trace fossils at some horizons indicates that life forms were plentiful during some stages of deposition.

Fossil evidence has been used in sedimentological interpretations of lake-water geochemistry on a general level, as noted above. Trace fossils have also been successfully used to deduce the nature of sediments soon after or during deposition. For example, burrows with indistinct walls were probably subaqueous (Bromley and Asgaard 1972), while vertebrate tracks associated with raindrop imprints were formed subaerially.

Few trace fossils are definitively marine or nonmarine, but micropaleontology has been used in recent studies to make such distinctions. Marine conditions, or at least marine source waters, have been inferred in the English Triassic on the basis of a microplankton assemblage (Warrington 1967, 1970). Palynological studies have been used in a number of basins for definitive age-dating in these otherwise poorly dated sediments. Palynological evidence of conditions in the immediate depositional environment is nebulous because of the wide distribution potential of pollen, but climates have been inferred from such data (Cornet and Traverse 1975).

Robbins (1982) has suggested that certain types of amorphous organic materials can be correlated with certain types of algae, and that the type of algae can be used to infer general water chemistry: blue-green algae imply nutrient-rich, phosphatic waters and closed-basin conditions, while green algae probably imply cleaner, clearer waters and some through-drainage.

Petrography and Grain-Size Analysis

Few recent sedimentological studies of the Triassic rift basins have made extensive use of either thin-section petrography or grain-size analysis techniques. Glaeser (1966) and Krynine (1950) presented detailed, grain-size studies; but the ratio of paleogeographic conclusions to generated data in those instances was significantly lower than can now be produced by studies of sedimentary structures. Krynine showed that there was little mixing of adjacent suites of trace minerals as they were dispersed laterally across the Hartford basin, and he deduced that drainage was primarily east–west rather than longitudinal (north–south). Hubert et al. (1978) have since corroborated this conclusion by measuring and mapping crossbed azimuths, but their data base was much more definitive and probably much more easily obtained than Krynine's.

Recently, Vetter and Brakenridge (1986) have proposed to turn this sequence around again and to test the paleogeographic reconstructions derived from crossbedding by means of petrographic techniques. In part, this study is also an attempt to produce a framework for the subsurface stratigraphic correlation of petrographic horizons in drill holes, where it is difficult to assess sedimentary structures and other criteria that are easily obtainable from outcrops.

Bluck's (1965) analysis of grain-size trends on ancient alluvial fans showed that the rate of decrease of grain sizes slows abruptly at a specific distance from the apex of the fan; but there is nothing to relate this to in the modern, where the rate of grain-size decrease is not a universally constant function of the alluvial fans' slopes. In modern fans, too much variation occurs in the slopes and rates of clast attrition; however, Bluck's 1967 study made definitive use of the general correlation between clast size and bed thickness to argue convincingly for a significantly faulted basin margin, in the absence of more direct evidence (see chapter 5).

Most recent petrographic studies have addressed diagenetic processes. For example, Hubert and Reed (1978) used thin-section data to document the sources of red coloration and diagenetic sequence in Triassic-Jurassic rocks. An exception is the study of Weddle and Hubert (1983), which combined the distribution of petrographic characteristics of rocks with paleoflow patterns and facies distributions to argue against the broad-terrane hypothesis in Connecticut and Massachusetts. Weddle and Hubert were able to show that, although Triassic sediments probably extended beyond their present limits in the Newark and Hartford/Deerfield basins, the two general areas were not connected as a continuous, broad area of subsidence and sedimentation. They reported that the Pomperaug outlier, however, contains a mineralogical suite, paleoflow vectors, and facies and grain-size trends that suggest that this small subbasin was continuous with the main Hartford basin during Late Triassic–Early Jurassic time.

Features of Modern Environments that Are Preserved in Rift-Basin Deposits

This section presents a brief review of the sedimentary characteristics from modern environments that have been reported from Triassic-Jurassic rift-basin sedimentary rocks, and that therefore are thought to have the best preservation and/or recognition potential. Most facies models are a combination of characteristics from modern environments and the ancient strata that are interpreted as having been deposited in similar environments, but a purely modern–ancient comparison will be made here. While most of the rift-basin deposits conform reasonably well to accepted facies models, numerous features of modern environments do not fit with patterns found in the rift basins. These features either have not been preserved or have not been recognized in the Triassic-Jurassic strata.

Alluvial-fan Deposits

The sedimentology of alluvial fan environments is relatively well understood. The processes and deposits have been well documented (as in Bull
1972), and ancient alluvial-fan deposits are easily recognized. Characteristic
sedimentary features common to both modern and ancient alluvial fans
include the following: radial paleoflow; rapid clast-size decrease downslope;
debris-flows, sheetflood (both midfan and fan-toe) and stream-channel deposits; and caliche profiles.

A number of additional modern features have not been preserved or have
gone unrecognized in the Triassic–Jurassic sedimentary rocks. These include
fan-head entrenchment, fan-slope gullying (except possibly in the case of
Bluck's 1967 description of fan deposits in Scotland), and other features
indicative of pauses in deposition (such as desert pavements and varnishes).
Moreover, the levees or lateral ridges common to debris flows have not been
reported. Many of these features are characteristic of the fan-apex area,
which probably has the least preservation potential. This area is also
volumetrically the least-abundant subfacies, so that chances of finding such
deposits are small. Others of the absent features are common only to
modern fans in true arid deserts and may not have developed on semiarid
Triassic–Jurassic fans. Additionally, the sizes of some of these features may be
significantly larger than the scale of most outcrops, leading to problems
of recognition.

Two features of modern fans have ambiguous representation in the rift
basins. These are sieve deposits and mud flows. Mud flows are the finer-
grained subcategory of debris flows, and their distinctiveness is commonly
lost among the other muddy interbeds in fan deposits. Sieve deposits,
however, require special clast sizes and shapes (and therefore a specific type
of source area) in order to form; thus, they probably are rare in the first
place, and—being most common to the fan-apex area—they may often have
been destroyed.

Fluvial Deposits

Studies of modern rivers and their deposits are too numerous to allow specific comparisons with the Triassic–Jurassic sedimentary rocks to be made
here for each and every type. Discussion is therefore limited to characteristics
of the ancient deposits that are common to most modern rivers of braided or
meandering habit.

Braided rivers are well represented in the rift basins, but the recognition of
their specific features has usually been an exercise in ambiguity. The most

characteristic features of modern braided rivers are complex channel patterns and ephemeral bars of numerous types. The presence of bars in the Triassic can usually be inferred by the presence of smaller-scale sedimentary structures such as avalanche slipfaces and horizontally bedded upstream slopes (Hubert and Forlenza 1987; Steel and Thompson 1983), but the reconstruction of specific bar types or positions within channels has been qualitative.

Conspicuously rare in reports of Triassic braided rivers have been sedimentary structures indicative of fluctuating discharge and water levels. Except for brief mention in the reports of Hubert and Forlenza (1987) on the Fundy basin, and Ramos and Sopeña (1983) in eastern Spain, features such as reactivation surfaces, low-stage dissection of high-stage bed forms, and deposition of finer-grained, low-stage deposits among the coarse-grained bars have not been described. Steel and Thompson (1983), however, have suggested that alternate layers of clast-supported and matrix-supported gravels in England may be attributable to processes of fluctuating discharge, while Tucker (1977, p. 174) inferred "discontinuous recession of flood waters" from an outcrop showing "the stepped [channel] margin of a conglomerate."

Interestingly, the lateral continuity of a number of inferred braided-river deposits has been ascribed to a lateral migration of the paleochannels (Stevens and Hubert 1980; Bluck 1967). Lateral migration is a phenomenon that is usually associated with models of meandering rivers, whereas the braided-river model most often accounts for lateral shifts only by avulsion. This intermixing of processes of the two models is made necessary by data evident in the outcrops. Such an occurrence would seem improbable only to geologists who think of facies models in terms of rigidly separate categories. Similarly, fining-upward grain-size trends are usually thought of as characteristic of meandering fluvial systems, but they have been reported from a number of rift-basin fluvial sequences that are interpreted, on the basis of other solid evidence, as being braided in origin.

Deposits of meandering rivers are relatively rare in the rift-basin strata. They are usually recognized on the basis of the presence of fining-upward grain sizes and/or lateral accretion surfaces preserved by lateral migration—both well-documented features in modern analogues. Few of the other features associated with modern meandering rivers have been reported, including abandoned channels and clay plugs, stacked meander belts, cutbank slumps, levee and splay deposits, and chute channels and bars.

The absence of most of these features can probably be attributed to the small size and volumetric insignificance of meandering fluvial systems in the rift basins. Some of these characteristics, however, are also missing in many other reconstructions of ancient meandering depositional environments, and this may be due in part to problems of preservation.

Studies of the characteristics of overbank deposits in both modern and Triassic environments have been generally neglected. Although features such as roots, paleosols, desiccation cracks, and small floodplain lakes have been reported from many of the sedimentary rocks in the rift basins, few studies have focused on this muddy lithofacies.

Paludal Deposits

Coal-bearing paludal deposits are relatively rare in the Triassic rift basins, and for the most part their sedimentological characteristics have not been studied in detail. Moreover, current ideas and facies models of coal deposition are in a state of flux. McCabe (1984) has suggested that many of the environments of modern peat deposition that have been studied in detail are flooded with clastic debris too often to be capable of producing thick coals analogous to those of the ancient. Although there is little published basis for comparison of the Triassic rift-basin coals with modern environments—and it will not be attempted here—the generally high rates of subsidence and sedimentation (hundreds of meters per million years) may be one of the reasons that coals did not form in many of these basins.

Lacustrine Deposits

Lacustrine sediments generally have good preservation potential, and most of the sedimentary structures and lithologies that are present in modern environments have been recognized in ancient strata. Modern clastic features such as slumps on the steeper depositional slopes, oscillation and current ripples, and turbidites are represented in many or all of the Triassic–Jurassic lacustrine strata. Modern chemical features such as oolites, oncolites, and numerous types of evaporites and other chemical-precipitation deposits are also present. Many of the lacustrine kerogen-bearing mudstones were produced in part by organic processes that are now being recognized and investigated more fully (Porter and Robbins 1981).

Perhaps the major difference between modern lacustrine sediments and strata of the Triassic–Jurassic that are interpreted as being lacustrine is the ancient record of variable lacustrine conditions. The time element is a major factor in lacustrine environments, where sedimentation rates are relatively slow and where the environment is subject to rapidly changing climatic and/or tectonic controls. This, combined with good preservation, results in a relatively complete sedimentary record with significant variation over short stratigraphic intervals. In the ancient, such variation has been recognized in the form of cycles, but these cycles are not generally apparent in many of the modern environments that have been studied. Moreover, although most of the Triassic–Jurassic lacustrine sedimentary cycles are assumed to have been caused by "some regional control, presumably cyclically changing climate" (Olsen 1986, p. 847), no asymmetry is yet apparent in the proposed mechanisms for climatic variation, and thus the reason for the asymmetry in the first-order sedimentation units remains unexplained.

Playa Deposits

Modern playas have only recently come to be considered important depositional environments. As a result, they have only been studied closely within the last fifteen or twenty years. Most analogies involving rift-basin playa deposits have been to playas of Holocene, desert, Basin and Range basins. Almost all of the clastic sedimentary structures—such as graded, fan-toe sheetflood deposits, ripple marks, desiccation cracks, and vertebrate tracks of the North American Triassic–Jurassic basins—have counterparts in these modern playa environments.

The evaporite/brine geochemistry of the ancient, however, often does not correlate with that of modern environments. Many of the Basin and Range basins contain thick evaporite sequences (often thick halites), arranged in concentric patterns with the most soluble evaporites concentrated in the lowest or central parts of the basin. Such concentrations and bull's-eye patterns have not yet been documented in the ancient. Van Houten (1964) has suggested that clastic Lockatong cycles predominate at the margins of the Newark basin, whereas chemical cycles are more common near the center, but there is no clear indication that the evaporite mineralogy itself varies laterally.

Many of the rift-basin playa deposits are overwhelmingly clastic in composition, with only scattered pseudomorphs of halite or other soluble minerals left to indicate that hypersaline conditions existed. Moreover, many of the ancient evaporite minerals are more complex than are the sulfate-to-chloride concentric patterns of the modern playas, and significant variation occurs in the suites of evaporites between the different rift basins—and even, at times, between different strata within a single basin. Possible modern analogues for rift-basin brine chemistries, other than those of the Basin and Range desert playas, might include the playa waters of the semiarid plains of western Canada, described by Last (1984). Most of these playas are dominated by sulfate, sodium, and magnesium ions, and the detailed chemistry of these brines varies from playa to playa, as well as seasonally and even daily. These playas also support abundant zooplankton.

Marine-influenced Deposits

Two basic divisions of the Triassic rift-basin facies of Europe and North Africa are covered under the heading of marine-influenced deposits: primarily subaqueous salina deposits; and primarily subaerial sabkha-type deposits. Salinas appear to have formed in low-lying, easily flooded graben areas, where thick halites and (locally) potash minerals were precipitated. Sulfates and sulfate-bearing mudstones are more uniformly present in both the graben and intergraben areas. Sedimentary structures in the Triassic halites are rarely preserved or observed—probably as a result of common recrystallization— but locally the textures of floor-nucleated crystals can be compared to halites forming in modern shallow salinas. Crossbedded halites indicative of currents or storms have not yet been reported.

Concentric patterns of evaporites develop in modern salinas as they do in many playas, and it is common to find sulfate-bearing mudstones near the margins of the Triassic grabens as facies equivalents to central-basin halites. In Morocco, later-stage potash evaporites can be found in the central parts of some of the basins. These late-stage minerals are rare in modern environments.

Triassic sabkha-type deposits are found both within and surrounding the salina halites in the grabens. These sulfate-rich mudstones are similar to modern sabkha deposits in many respects—primarily in the volumetric dominance of muddy matrix material over evaporite minerals, and in the disruption of the muddy sediments by syndepositional and early diagenetic growth of evaporites (usually gypsum/anhydrite).

The ancient sabkhas also display many dissimilarities to the modern analogues. One of these is the vast, continental-scale, lateral extent of Triassic sabkha environments. Warren and Kendall (1985) report modern sabkha environments of 10 to 15 kilometers in width and 150 kilometers in length parallel to shorelines, but there was no definitive Triassic shoreline in Europe and North Africa. Sabkha-type environments developed on the margins of graben-fill salinas, occasionally within the grabens, and across the broad, eroded, low-relief areas between grabens, all of which were apparently subject to periodic inundations.

Thus, these ancient sabkhas do not display the shoaling-upward profiles or beach ridges found in many modern prograding-shoreline sabkhas. For the same reason, equivalent "offshore" carbonate or clastic facies do not exist. There is also little evidence found in the ancient mudstones of the algal mats that are diagnostic of modern sabkha shorelines. Although this may be in part a matter of lack of preservation of organic material in the oxidized, red Triassic deposits, it may equally be due to the ephemeral nature of the shorelines in comparison to those of modern sabkha environments.

Few studies have been made of the coarser clastic lithofacies that are sometimes associated with these marine-influenced deposits. Ireland et al. (1978) recognized tidal channels and intertidal mud flats in some of the muddy, silty, and (rarely) sandy facies that grade up into the salt beds in England; but in general, few definitive, tidally produced structures such as sigmoidal tidal bundles have been described. The general absence of sandy detritus and the possibility of low or no tidal influence in many of the basins may be a contributing factor to the absence of such structures. The lack of recognition could also be important, especially if tidally produced structures are expected to be commonly found in the deposits of embayed shallow seaways (Kreisa and Moiola 1986), as the Late Triassic environments of Europe and North Africa may have been (albeit episodically and ephemerally).

Eolian Deposits

Few rift-basin eolian deposits have been described in detail because few exist. Thus, a limited suite of sedimentary characteristics exist that are common to both the Triassic–Jurassic strata and the modern heterogeneous eolian environments. Most small-scale sedimentary structures, however, are analogous between the modern and ancient rift-basin environments. These include grain-flow and grain-fall dune foresets, relatively well-rounded sand grains, and eolian ripples.

Missing from descriptions of the rift-basin eolianites of North America are large, avalanche-face slumps or garland structures, but these are usually rare in arid eolian systems, being more common in damp/dry eolian deposits (Krystinik 1986). These features do occur locally in the dome-shaped dunes of central England (Thompson 1969).

Deflation lags in between successive dune beds and in the interdune areas are common to both modern and ancient deposits. Water-table deflation surfaces without pebble lags, however, have rarely been described in the rift-basin environments. Moreover, most of the interdune deposits of the Triassic apparently do not display the rooting, burrowing, and eolian rippling that characterize many modern interdune environments.

Scientific Methods for a Historical Science

As was noted in the preface to this book, there is a general tendency among scientists to feel condescension toward nonnumerical, historical sciences, including geology. As early as 1857, James Hall was writing that "I must protest against the narrow view which has been expressed by some, that Natural History is not a science, and who perhaps believe that mathematical formulae lie at the basis of all science and that these must be used in all their minutiae to reach any valuable result" (Hall 1882, p. 31).

If we accept that geology cannot always be deductive and quantitative, what rigorous, scientific methods can be applied to this historical science in which inductive reasoning and observation are the main tools of investigation? Gould (1986) has listed a number of scientific methodologies that can be applied to historical sciences, and these are paraphrased below (modified freely for application to sedimentology). Examples from the Triassic–Jurassic rift-basins are given for each method listed.

The Present as Key to the Past

The time-honored geological tenet that the present is the key to the past has come under recent scrutiny with the debate over uniformitarian versus actualistic processes, but the principle remains a valid one despite probable changes in the scale of its application. It is a major concept about the nature of time and additive processes, by which presently observable events (whether day-to-day or hundred-year high-energy events) produce definable deposits that accumulate over time—deposits that can be observed as a cumulative record in ancient strata.

The methodology of this concept is common knowledge: the processes and deposits of present environments can be studied, and the characteristics of the deposits can be compared to ancient strata; similarities can then be used to infer similar depositional processes that took place in similar environments in the geological past. This method is the basis for sedimentary facies models, and has been invaluable in all aspects of sedimentological study of the rift basins. A prime example of this is the reconstruction of the rift-basin alluvial fans.

An example of an application of this method that did *not* work is the early analogy that was drawn between the estuarine Bay of Fundy and the North American rift basins (see Part I). The principal reason for this failure was the inadequacy of both the modern and the ancient data bases that were used for the comparisons.

Categorization and Relationships

Faced with a vast data base for which no previous experience (and thus no frame of reference for interpretation) exists, the geologist must begin by classifying the different data in order to recognize the different categories that are applicable, and the similarities and differences between them. This is the primary purpose and important function of stratigraphic nomenclature, although nomenclature has (unfortunately) often become an end in itself. Lithologic and/or stratigraphic mapping is an example of this preliminary sorting and categorizing of the voluminous data present in the geological record. Where the initial sorting was done on the basis of faulty premises—as was Emerson's early map of the northern half of the Hartford basin—subsequent interpretations of the relationships of the data categories, as well as the final model of paleogeography, were strained.

Once categories are erected, the second step of interpretation can be taken. Thus, where data have been reasonably categorized (there are few absolutely correct categorizations), the relationships among them can be deduced. Different categories can often be recognized as successive stages of as-yet unknown processes—processes that may be reconstructed from properly categorized data sets with empirically known relationships. The models of Triassic–Jurassic lacustrine cycles put together by Van Houten (1962, 1964) and Hubert, Reed, and Carey (1976) prior to the emergence of detailed sedimentological knowledge of modern lacustrine environments are excellent examples of the application of this method to geological problems. In these cases, the rocks provided clues that allowed reconstruction of ancient environments even though modern analogues were not yet well known or were imperfectly analogous.

A variation of this method involves the recognition of common patterns despite overall complexity. Patterns of vertical or lateral grain-size distribution, such as in fining-upward fluvial sandstones, tectonic fining-upward of formations, or radial decrease in grain sizes on alluvial fans, illustrate common patterns that can be isolated from the prevailing complexity of rift-basin sedimentary rocks.

Gould also suggests that a certain amount of testability can be applied in the use of this method, by *postdicting* (versus *predicting*) the occurrence of patterns—that is, by extrapolating patterns from known to unknown areas. Geologically, this can be done by extrapolating facies from known to unknown outcrops or from outcrops to the subsurface, using geological principles and known trends of facies changes.

The proof is in the finding of the postdicted patterns in the tested areas. This is a common geological technique in economic exploration (where, for example, testing for predicted oil reservoirs is performed by drilling into the unknown subsurface), and it may soon be applied to the North American rift basins in the search for hydrocarbons. It is indicative of the complexity of geology—and perhaps the current state of the science—that of the predicted hydrocarbon pay zones drilled by wildcat wells, only 10 percent prove out into producing wells.

Anomalies as Legacies of Past History

Anomalies are often apparent in a geological data set. These are data that do not fit into the overall categories that have been erected. Such data (if real) may be inherited, reflecting a previous set of geological conditions rather than the ones that are being reconstructed to explain the majority of the data. Gould offers the example of certain apparently anomalous distributions of fossils as data that had fit a coherent pattern prior to continental breakup, but that are now widely scattered without apparent reason unless the processes of plate tectonics are taken into consideration.

In the Triassic–Jurassic rift basins, an example of an apparent anomaly is the orientation of basin axes that are commonly oblique to the inferred Triassic extensional stresses. On a homogeneous basement, one would expect normal and listric faults to form perpendicular to the extensional stresses. The apparently anomalous structural orientations of the rift basins are resolved when the strong control of the preexisting structures and structural fabric of the basement rocks are taken into account.

Sedimentologically, inherited basement structure explains the end-of-the-basin position of the alluvial fans in the rift basins of the High Atlas of Morocco. These basins are aligned along a Paleozoic suture zone (Van Houten 1977) that was locally intruded by a late Paleozoic plutonic complex (Vogel et al. 1976). Triassic rifting was preferentially aligned along the axis of the suture, but the plutonic massif did not subside along with the grabens. Instead, it remained as a high-standing sediment source that shed alluvial-fan conglomerates east and west into the end regions of the grabens.

Consilient Theories

Gould has not proposed consilience (the concurrence or accordance of different and unassociated inferential results) as an enumerated fourth method, but he has described it and discussed its usefulness, and it is included here. The concept is that there should be a single, comprehensive (preferably uncomplicated) theory that explains all or most facets of the evidence, even though each piece of data does not prove the theory in and of itself.

A series of consilient theories have been proposed for the Hartford basin, including Rogers's early concept of major axial fluvial systems, the early ideas of estuarine environments, and the more recent theories of enclosed basins and alluvial fans (see Part I). Each successive theory took into account a wider and more detailed data base, with fewer exceptions for perceived anomalies. Barrell's (1915) formulation of a consilient theory of semiarid alluvial-fan and fluvial deposition was much less complex and accounted for data more easily than Emerson's contemporaneous (1917) concept of mixed estuarine, glacial, shoreline, and central raised-mud-flat environments.

Conclusion

The use of these four generalized techniques and their subsidiaries as methodologies for sedimentological study is common to the science of geology, although they are seldom enumerated. The point to be made here is that they are scientific methods in their own right and ought to be taught, thought of, and used as such.

In conjunction with this (if my background can be taken as typical), education in sedimentology consists primarily of learning the detailed characteristics of modern environments and ancient deposits. This is proper enough, but seldom is a course designed specifically to teach how to make rigorous comparisons between the two; in fact, the distinction between them is commonly blurred. As a result, that particular aspect of one's education is usually left to the process of trial and error and can result in some dubious interpretations of outcrops until one develops the capability outside the classroom.

It is impossible to overestimate the importance to a sedimentologist's education of wide-ranging, first-hand field experience with the unlimited combinations and permutations found within modern sedimentary environments and ancient deposits. In a science that operates in large part by analogy, it is restrictive to have too small a personal stock of potential analogues, and reading about facies models is amazingly less effective in imparting an appreciation for them than is field experience. While a general feel for a facies model or sedimentary sequence can be gained through the literature, the sedimentological concepts are most forcefully internalized, and become an active part of the geologist's capabilities, when they are experienced in situ. Field experience also keeps a geologist aware of ambiguities and the different possibilities in the interpretation of a suite of rocks. It helps one to maintain a flexible approach to the application of facies models, and when undertaken with an open mind, it helps one to maintain a certain degree of humility.

References, Part I

Allen, J. R. L. 1965. A review of the origin and characteristics of recent alluvial sediments. *Sedimentology* 5:89–191.

American Journal of Science. 1843. [Anonymous review of Percival's 1842 *Report on the Geology of the State of Connecticut.*] *Am. J. Sci.* 44:187–88.

Association of American Geologists and Naturalists. 1842. Minutes of the annual meeting. *Am. J. Sci.* 43:150–73.

Association of American Geologists and Naturalists. 1843. *Reports of the first, second, and third meetings of the Association of American Geologists and Naturalists, at Philadelphia, in 1840 and 1841, and at Boston in 1842: proceedings and transactions.* Boston: Gould, Kendall, and Lincoln.

Austin, J. A., Jr.; Uchupi, E.; Shaughnessy, D. R.; and Ballard, R. D. 1980. Geology of New England passive margin. *Bull. Am. Ass'n Petrol. Geol.* 64:501–26.

Bain, G. W. 1932. The northern area of Connecticut Valley Triassic. *Am. J. Sci.* 230:57–77.

———. 1957a. Triassic age rift structure in eastern North America. *Trans. N.Y. Acad. Sci.* Ser. 2, 19:489–502.

Bain, G. W., ed. 1957b. Guidebook: Geology of northern part—Connecticut Valley. In *49th New England Intercollegiate Geological Conference*, p. 1–26.

Barrell, J. 1906. Relative importance of continental, littoral, and marine sedimentation. *J. Geol.* 14: Part I [untitled], 316–56; Part II [untitled], 430–57; Part III, Mud-cracks as a criteria of continental sedimentation, 524–68.

———. 1908. Relations between climate and terrestrial deposits. *J. Geol.* 16: Part I, Relations of sediments to regions of erosion, 159–90; Part II, Relations of sediments to regions of deposition, 255–95; Part III, Relations of climate to stream transportation, 363–84.

———. 1915. *Central Connecticut in the geologic past.* State Geol. and Nat. Hist. Survey of Conn. Bull. 23.

———. 1917. Rhythms and the measurements of geologic time. *Bull. Geol. Soc. Am.* 28:745–904.

Barrell, J., and Loughlin, G. F. 1910. *Lithology of Connecticut.* State Geol. and Nat. Hist. Survey of Conn. Bull. 13.

Bell, M. 1985. *The face of Connecticut: people, geology, and the land.* State Geol. and Nat. Hist. Survey of Conn. Bull. 110.

Berner, R. A. 1969. Goethite stability and the origin of red beds. *Geochim. et Cosmochim. Acta* 33:267–73.

Bouma, A. H. 1962. *Sedimentology of some flysch deposits: a graphic approach to facies interpretation.* Amsterdam: Elsevier.

Brewer, R. 1964. *Fabric and mineral analysis of soils.* New York: John Wiley.

Brophy, G. B.; Foose, R. M.; Shaw, F. C.; and Szekely, T. S. 1967. Triassic geologic features in the Connecticut Valley near Amherst, Massachusetts. In *Guidebook for field trips in the Connecticut Valley of Massachusetts: 59th New England Intercollegiate Geological Conference,* ed. P. Robinson. p. 61–72.

Burchfield, J. D. 1975. *Lord Kelvin and the age of the Earth.* New York: Science History Pub.

Byrnes, J. B., and Horne, J. C. 1974. Alluvial fan to marine facies of Connecticut Valley Triassic [abstract]. *Am. Ass'n Petrol. Geol./Soc. Econ. Paleont. Mineral. Annual Meeting Abstracts* 1:15.

Chandler, W. E. 1978. *Graben mechanics at the junction of the Hartford and Deerfield basins of the Connecticut Valley, Massachusetts.* Dept. of Geol. and Geog., Univ. of Mass., Contribution 33.

Chapman, R. W. 1965. *Stratigraphy and petrology of the Hampden Basalt in central Connecticut.* State Geol. and Nat. Hist. Survey of Conn. Report of Invest. 3.

Colbert, E. H., and Gregory, J. T. 1957. Correlation of continental Triassic sediments by vertebrate fossils. In Correlation of the Triassic formations of North America, exclusive of Canada, ed. J. B. Reeside. *Bull. Geol. Soc. Am.* 68:1456–67.

Colton, R. B., and Hartshorn, J. H. 1966. *Bedrock geology map of the West Springfield quadrangle, Massachusetts and Connecticut.* U.S. Geol. Survey Map GQ537.

Cooper, T. 1822. On volcanoes and volcanic substances, with a particular reference to the origin of the rocks of the floetz trap formation. *Am. J. Sci.* 4:205–43.

Cornet, B., and Traverse, A. 1975. Palynological contributions to the chronology and stratigraphy of the Hartford basin in Connecticut and Massachusetts. *Geoscience and Man* 11:1–33.

Cornet, B.; Traverse, A.; and McDonald, N. G. 1973. Fossil spores, pollen, and fishes from Connecticut indicate Early Jurassic age for part of the Newark Group. *Science* 182:1243–46.

Crosby, W. O. 1885. Color of soils. *Proc. Boston Soc. Nat. Hist.* 23:219–22.

———. 1891. On the contrast in color of the soils of high and low latitudes. *Am. Geologist* 8:72–82.

Dana, J. D. 1863. *Manual of geology.* 1st ed. Philadelphia: T. Bliss.

———. 1873. On some results of the Earth's contraction from cooling, including a discussion of the origin of mountains and the nature of the Earth's interior. *Am. J. Sci.* 5:423–43.

———. 1879. [Critical review of 1879 article by I. C. Russell.] *Am. J. Sci.* 17:328–30.

———. 1883. [Review of the 1882 *Annual Report of the State Geologist of New Jersey* by G. H. Cook.] *Am. J. Sci.* 25:383–86.

———. 1891a. Some of the features of non-volcanic igneous ejections, as illustrated in the four "rocks" of the New Haven region, West Rock, Pine Rock, Mill Rock, and East Rock. *Am. J. Sci.* 42:79–110.

———. 1891b. On Percival's map of the Jura–Trias trap belts of central Connecticut, with observations on the upturning, or mountain-making disturbance, of the formation. *Am. J. Sci.* 42:439–47.

———. 1896. *Manual of geology.* 4th ed. New York: American Book.

Davis, W. M., 1882. Brief notice of observations on the Triassic trap rocks of Massachusetts, Connecticut, and New Jersey. *Am. J. Sci.* 24:345–49.

———. 1886. The structure of the Triassic formation of the Connecticut Valley. *Am. J. Sci.* 32:342–52.

———. 1888. The structure of the Triassic formation of the Connecticut Valley. In *7th Annual Report, U.S. Geol. Survey,* pp. 457–90.

———. 1896. The quarries in the lava beds at Meriden, Connecticut. *Am. J. Sci.* 1:1–13.

————. 1898. The Triassic formation of Connecticut. In *18th U.S. Geol. Survey, Annual Report,* Part 2, p. 1-192.

Davis, W. M., and Griswold, L. S. 1894. Eastern boundary of the Connecticut Triassic. *Bull. Geol. Soc. Am.* 5:515-30.

Davis, W. M., and Loper, S. W. 1891. Two belts of fossiliferous black shale in the Triassic formation of Connecticut. *Bull. Geol. Soc. Am.* 2:415-30.

de Boer, J. 1967. Paleomagnetic-tectonic study of Mesozoic dike swarms in the Appalachians. *J. Geophys. Research* 72:2237-50.

————. 1968*a*. Late Triassic volcanism in the Connecticut Valley and related structure. In *Guidebook for field trips in Connecticut: 60th New England Intercollegiate Geological Conference,* ed. P. M. Orville, Guidebook No. 2, trip C-5, p. 1-12.

————. 1968*b*. Paleomagnetic differentiation and correlation of the Late Triassic volcanic rocks in the central Appalachians (with special reference to the Connecticut Valley). *Bull. Geol. Soc. Am.* 79:609-26.

Demicco, R. V., and Gierlowski-Kordesch, E. G. 1986. Facies sequences of a semi-arid closed basin: the Lower Jurassic East Berlin Formation of the Hartford basin, New England, USA. *Sedimentology* 33:107-18.

Dewey, J. F., and Bird, J. M. 1970. Mountain belts and the new global tectonics. *J. Geophys. Research* 75:2625-47.

Dorsey, G. E. 1926. The origin of the color of red beds. *J. Geol.* 34:131-43.

Eastman, C. R. 1911. *Triassic fishes of Connecticut.* State Geol. and Nat. Hist. Survey of Conn. Bull. 18.

Eiseley, L. 1958. *Darwin's century: evolution and the men who discovered it.* Reprint 1961. Garden City, NY: Doubleday.

Emerson, B. K. 1891. On the Triassic of Massachusetts. *Bull. Geol. Soc. Am.* 2:451-56.

————. 1898*a*. *Holyoke folio—Massachusetts–Connecticut.* U.S. Geol. Survey Folio 50.

————. 1898*b*. *Geology of Old Hampshire County, Massachusetts.* U.S. Geol. Survey Monograph 29.

————. 1917. *Geology of Massachusetts and Rhode Island.* U.S. Geol. Survey Bull. 597.

Fenton, C. L., and Fenton, M. A. 1952. *Giants of geology.* Garden City, N.Y.: Doubleday.

Foose, R. M.; Rytuba, J. J.; and Sheridan, M. F. 1968. Volcanic plugs in the Connecticut Valley Triassic near Mt. Tom, Massachusetts. *Bull. Geol. Soc. Am.* 79:1655-62.

Foye, W. G. 1922. Origin of the Triassic trough of Connecticut. *J. Geol.* 30:690-99.

————. 1924. Pillow structure in the Triassic basalts of Connecticut. *Bull. Geol. Soc. Am.* 35:329-46.

Froelich, A. J., and Olsen, P. E. 1984. *Newark Supergroup, a revision of the Newark Group in eastern North America.* U.S. Geol. Survey Bull. 1537-A, p. 55-58.

Gierlowski-Kordesch, E. G., and Demicco, R. V. 1983. Playa deposition in a closed basin: East Berlin Formation [abstract]. *Geol. Soc. Am. Abstracts with Programs* 15:122.

Glock, W. S. 1927. The significance of red color in sediments: discussion and communications. *Am. J. Sci.* 14:155-56.

Gould, S. J. 1986. Evolution and the triumph of homology, or why history matters. *Am. Scientist* 74:60-69.

Gray, N. H. 1982. Mesozoic volcanism in north-central Connecticut. In *Guidebook for field trips in Connecticut and south-central Massachusetts,* ed. R. Joesten and S. S. Quarrier, p. 173-90. State Geol. and Nat. Hist. Survey of Conn. Guidebook 5.

Greene, M. T. 1982. *Geology in the nineteenth century.* Ithaca, N.Y.: Cornell Univ. Press.

Hall, J. 1882. [Presidential address to the American Association for the Advancement of Science, written in 1857.] *Proc. Am. Ass'n Adv. Sci.* 31:29-71.

Hallam, A. 1983. *Great geological controversies.* London: Oxford Univ. Press.

Hand, B. M.; Wessel, J. M.; and Hayes, M. O. 1969. Antidunes in the Mount Toby Conglomerate (Triassic), Massachusetts. *J. Sed. Petrol.* 39:1310-16.

Hardie, L. A.; Smoot, J. P.; and Eugster, H. P. 1978. Saline lakes and their deposits, a sedimentological approach. In *Modern and ancient lake sediments,* ed. A. Matter and M. E. Tucker, p. 7-41. Int'l Ass'n Sedimentologists Special Pub. 2.

Heezen, B. C. 1960. The rift in the ocean floor. *Scientific American* 203:98-110.

High, L. R., Jr. 1985. Evaluation of potential hydrocarbon sources in lacustrine facies of Newark Supergroup, eastern United States [abstract]. *Bull. Am. Ass'n Petrol. Geol.* 69:265.

Hitchcock, E. 1818. Remarks on the geology and mineralogy of a section of Massachusetts on Connecticut River. *Am. J. Sci.* 1:105-16, 436-39.

——. 1823, 1824. A sketch of the geology, mineralogy, and scenery of the regions contiguous to the River Connecticut: with a geological map and drawings of organic remains. *Am. J. Sci.* 6: Part I, Geology, 1-86; Part II, Mineralogy, 201-36. *Am. J. Sci.* 7: Part III, Scenery, 1-30.

——. 1828. Miscellaneous notices of mineral localities, with geological remarks. *Am. J. Sci.* 14:215-30.

——. 1836. Ornithichnology—description of the foot marks of birds (ornithichnites) on New Red Sandstone in Massachusetts. *Am. J. Sci.* 29:307-40.

——. 1841. *Final report on the geology of Massachusetts.* 2 vols. Northampton: J. H. Butler.

——. 1843. Description of several species of fossil plants from the New Red Sandstone formation of Connecticut and Massachusetts. In *Reports of the first, second and third meetings of the Association of American Geologists and Naturalists,* p. 294-96. Boston: Gould, Kendall, and Lincoln.

——. 1844. *Geological map of Massachusetts, 1:316,800, and explanation of the geological map, attached to the topographical map of Massachusetts.* Boston: C. Hickling.

——. 1851. *The religion of geology and its connected sciences.* London: William Collins.

——. 1858. *Report on the sandstone of the Connecticut Valley, especially its fossil footmarks.* Boston: William White.

——. 1865. *Supplement to the ichnology of New England.* Boston: Wright and Potter.

Hobbs, W. H. 1901. The Newark System of the Pomperaug Valley, Connecticut. In *U.S. Geological Survey, 21st Annual Report,* Part III, pp. 7-160.

——. 1902. Former extent of the Newark System. *Bull. Geol. Soc. Am.* 13:139-48.

Hogg, S. E. 1982. Sheetfloods, sheetwash, sheetflow, or . . . ? *Earth-Science Reviews* 18:59-76.

Holmes, A. 1915. Radioactivity and the measurement of geologic time. *Proc. Geol. Ass'n* 26(5):289-309.

Hopkins, D. M., ed. 1967. *The Bering land bridge.* Stanford, Calif.: Stanford Univ. Press.

Hsü, K. J. 1973. The odyssey of geosyncline. In *Evolving concepts in sedimentology,* ed. R. N. Ginsburg, p. 66-93. Baltimore: Johns Hopkins Univ. Press.

Hubert, J. F. 1977. Paleosol caliche in the New Haven Arkose, Connecticut: record of semi-aridity in Late Triassic-Early Jurassic time. *Geology* 4:302-4.

——. 1978. Paleosol caliche in the New Haven Arkose, Newark Group, Connecticut. *Palaeogeogr., Palaeoclim., Palaeoecol.,* 24:151-68.

——. 1986. Personal communication.

Hubert, J. F.; Gilchrist, J. M.; and Reed, A. A. 1982. Jurassic redbeds of the Connecticut Valley: (1) brownstone of the Portland Formation; and (2) playa-playa lake-oligomictic lake model for parts of the East Berlin, Shuttle Meadow, and Portland Formations. In *Guidebook for field trips in Connecticut and south-central Massachusetts,* ed. R. Joesten and S. S. Quarrier, p. 103-41. State Geol. and Nat. Hist. Survey of Conn. Guidebook 5.

Hubert, J. F., and Hyde, M. G. 1982. Sheet-flow deposits of graded beds and mudstones on an alluvial sandflat–playa system: Upper Triassic Blomidon redbeds, St. Mary's Bay, Nova Scotia. *Sedimentology* 29:457–74.

Hubert, J. F., and Reed, A. A. 1978. Red-bed diagenesis in the East Berlin Formation, Newark Group, Connecticut Valley. *J. Sed. Petrol.* 48:175–84.

Hubert, J. F.; Reed, A. A.; and Carey, P. J. 1976. Paleogeography of the East Berlin Formation, Newark Group, Connecticut Valley. *Am. J. Sci.* 276:1183–1207.

Hubert, J. F.; Reed, A. A.; Dowdall, W. L.; and Gilchrist, J. M. 1978. *Guide to the redbeds of central Connecticut: 1978 Field Trip, eastern section of Soc. Econ. Paleont. Mineral.* Dept. of Geol. and Geog., Univ. of Mass., Contribution 32.

Huxley, T. H. 1887. The progress of science. Reprinted in *Selections from the essays of T. H. Huxley.* 1948. New York: Appleton-Century-Crofts.

Kay, M. 1944. Geosynclines in continental development. *Science* 99:461–62.

———. 1951. *North American geosynclines.* Geol. Soc. Am. Memoir 48.

Kaye, C. A. 1983. Discovery of a Late Triassic basin north of Boston and some implications as to post-Paleozoic tectonics in northeastern Massachusetts. *Am. J. Sci.* 283:1060–79.

Klein, G. de V. 1962. Triassic sedimentation, maritime provinces, Canada. *Bull. Geol. Soc. Am.* 73:1127–46.

———. 1963. Bay of Fundy intertidal zone sediments. *J. Sed. Petrol.* 33:844–54.

———. 1968. Sedimentology of Triassic rocks in the lower Connecticut Valley: In *Guidebook for field trips in Connecticut, 60th New England Intercollegiate Geological Conference,* ed. P. M. Orville. Guidebook No. 2, trip C-1, p. 1–19.

———. 1969. Deposition of Triassic sedimentary rocks in separate basins, eastern North America. *Bull. Geol. Soc. Am.* 80:1825–32.

Kotra, R. K.; Hatcher, P. G.; Spiker, E. C.; Romankiw, L. A.; Gottfried, R. M.; Pratt, L. M.; and Vuletich, A. K. 1985. Organic geochemical investigations of eastern U.S. Early Mesozoic basins [abstract]. *Bull. Am. Ass'n Petrol. Geol.* 69:1439.

Krynine, P. D. 1935. Arkose deposits in the humid tropics: a study of sedimentation in southern Mexico. *Am. J. Sci.* 29:353–63.

———. 1942. Differential sedimentation and its products during one complete geosynclinal cycle. Part 1, Vol. 2 of *Anales del Primer Congresso Panamericano de Ingenieria de Minas y Geologia* (Santiago, Chile), p. 536–61.

———. 1950. *Petrology, stratigraphy, and origin of the Triassic sedimentary rocks of Connecticut.* State Geol. and Nat. Hist. Survey of Conn. Bull. 73.

Kuenen, P. H., and Migliorini, C. I. 1950. Turbidity currents as a cause of graded bedding. *J. Geol.* 58:91–127.

Lawson, A. C. 1913. The petrographic designations of alluvial fan formations. *Bull. Univ. of Calif. Dept. of Geol.* 7:325–34.

Lehman, E. P. 1959. *The bedrock geology of the Middletown quadrangle.* State Geol. and Nat. Hist. Survey of Conn. Quad. Report 8.

LeTourneau, P. M. 1984. Alluvial fan development in the Lower Jurassic Portland Formation, Connecticut [abstract]. *Geol. Soc. Am. Abstracts with Programs* 16:46–47.

———. 1985*a.* The sedimentology and stratigraphy of the Lower Jurassic Portland Formation, central Connecticut. Master's thesis, Wesleyan University.

———. 1985*b.* Lithofacies analysis of the Portland Formation in central Connecticut [abstract]. *Geol. Soc. Am. Abstracts with Programs* 17:31.

———. 1985*c.* Alluvial fan development in the Lower Jurassic Portland Formation, central Connecticut—implications for tectonics and climate. In *Proceedings of the second U.S. Geological Survey workshop on the early Mesozoic basins of the eastern United States,* ed. G. R. Robinson and A. J. Froelich, p. 17–26. U.S. Geol. Survey Circular 946.

LeTourneau, P. M., and McDonald, N. G. 1985. The sedimentology, stratigraphy, and paleontology of the Lower Jurassic Portland Formation, Hartford basin, central Connecticut. In *Guidebook for field trips in Connecticut and adjacent areas of New York and Rhode Island, 77th New England Intercollegiate Geological Conference,* ed. R. J. Tracy. p. 353-91.

LeTourneau, P. M., and Smoot, J. P. 1985. Comparison of ancient and modern lake margin deposits from the Lower Jurassic Portland Formation, Connecticut, and Walker Lake, Nevada [abstract]. *Geol. Soc. Am. Abstracts with Programs* 17:31.

Lewis, J. V. 1914. Origin of pillow lavas. *Bull. Geol. Soc. Am.* 25:591-654.

Lindholm, R. C. 1978. Tectonic control of sedimentation in Triassic-Jurassic Culpeper basin, Virginia [abstract]. *Bull. Am. Ass'n Petrol. Geol.* 62:537.

Little, R. D. 1982. Lithified armored mud balls of the Lower Jurassic Turners Falls Sandstone, north-central Massachusetts. *J. Geol.* 90:203-7.

Longwell, C. R. 1922. Notes on the structure of the Triassic rocks in southern Connecticut. *Am. J. Sci.* 4:223-36.

———. 1928. The Triassic of Connecticut. *Am. J. Sci.* 16:259-63.

———. 1933. The Triassic belt of Massachusetts and Connecticut. In *16th International Geological Congress, Guidebook 1,* p. 93-118.

———. 1937. Sedimentation in relation to faulting. *Bull. Geol. Soc. Am.* 48:433-42.

Lull, R. S. 1915. *Triassic life of the Connecticut Valley.* State Geol. and Nat. Hist. Survey of Conn. Bull. 24.

Luttrell, G. 1985. Personal communication, U.S. Geological Survey Stratigraphic Names Committee.

Lyell, C. 1843. On the fossil foot-prints of birds and impressions of rain-drops in the valley of the Connecticut. *Am. J. Sci.* 45:394-97.

———. 1845. *Journal of a tour in North America, 1841-1842.* 2 vols. New York: Wiley and Putnam.

———. 1851. On fossil rain-marks of the Recent, Triassic, and Carboniferous periods. *Quart. J. Geol. Soc. London* 7:238-47.

McBride, E. 1973. Concepts of Appalachian basin sedimentation. In *Evolving concepts in sedimentology,* ed. R. N. Ginsburg. pp. 93-117. Baltimore: Johns Hopkins Univ. Press.

Maclure, W. 1809. Observations on the geology of the United States explanatory of a geological map. *Trans. Am. Phil. Soc.* 6(2):411-48. Republished (1817) in monograph form. Philadelphia: A. Small.

———. 1824. Miscellaneous remarks on the systematic arrangement of rocks, and on their probable origin, especially of the Secondary. *Am. J. Sci.* 7:261-64.

Manspeizer, W.; Puffer, J. H.; and Couzminer, H. L. 1978. Separation of Morocco and eastern North America: a Triassic-Liassic stratigraphic record. *Bull. Geol. Soc. Am.* 89:901-20.

Merrill, G. P. 1906. Contributions to the history of American geology. In *Report of the United States National Museum for 1904, No. 135,* p. 189-734. Reprinted (1906) by the Government Printing Office, Washington, D.C.

———. 1924. *The first one hundred years of American geology.* New Haven: Yale Univ. Press.

Miall, A. D. 1978. Fluvial sedimentology: an historical review. In *Fluvial sedimentology,* ed. A. D. Miall, p. 1-47. Can. Soc. Petrol. Geol. Memoir 5.

Middleton, G. V. 1965. Antidune crossbedding in a large flume. *J. Sedimentary Petrology* 35:922-27.

Middleton, G. V. 1978. Sedimentology-history. In *The encyclopedia of sedimentology,* ed. R. W. Fairbridge and J. Bourgeois, p. 707-12. Stroudsburg, Pa.: Dowden, Hutchinson, and Ross.

Nadon, G. C., and Middleton, G. V. 1984. Tectonic control of Triassic sedimentation in southern New Brunswick: local and regional implications. *Geology* 12:619-22.

Newberry, J. S. 1887. The fauna and flora of the Trias of New Jersey and the Connecticut Valley. *Trans. N.Y. Acad. Sci.* 6:124-28.

————. 1888. *Fossil fishes and fossil plants of the Triassic rocks of New Jersey and the Connecticut Valley.* U.S. Geol. Survey Monograph 14.

Ogburn, C. 1968. *The forging of our continent.* New York: American Heritage.

Olsen, P. E. 1978. On the use of the term *Newark* for Triassic and Early Jurassic rocks of eastern North America. *Newsletters on Stratigraphy* 7:90–95.

————. 1983. On the non-correlation of the Newark supergroup by fossil fishes: biogeographic, structural, and sedimentological implications [abstract]. *Geol. Soc. Am. Abstracts with Programs* 15:121.

————. 1985*a*. Significance of great lateral extent of thin units in Newark Supergroup (lower Mesozoic, eastern North America) [abstract]. *Bull. Am. Ass'n Petrol. Geol.* 69:1444.

————. 1985*b*. Distribution of organic matter–rich lacustrine rocks in the early Mesozoic Newark Supergroup. In *Proceedings of the second U.S. Geological Survey workshop on the early Mesozoic basins of the eastern United States,* ed. G. R. Robinson and A. J. Froelich, pp. 61–64. U.S. Geol. Survey Circular 946.

Olsen, P. E. 1986. A 40-million year lake record of early Mesozoic orbital climatic forcing. *Science* 234:842–48.

Olsen, P. E.; McCune, A. R.; and Thomson, K. S. 1982. Correlation of the early Mesozoic Newark Supergroup by vertebrates, principally fishes. *Am. J. Sci.* 282:1–44.

Parnell, J. 1983. Skeletal halites from the Jurassic of Massachusetts, and their significance. *Sedimentology* 30:711–15.

Percival, J. G. 1842. *Report on the geology of the state of Connecticut.* New Haven: Osborn and Baldwin.

Pettijohn, F. J. 1984. *Memoirs of an unrepentant field geologist.* Chicago: Univ. of Chicago Press.

Philpotts, A. R., and Martello, A. 1986. Diabase feeder dikes for the Mesozoic basalts in southern New England. *Am. J. Sci.* 286:105–26.

Platt, J. N., Jr. 1957. Sedimentary rocks of the Newark Group in the Cherry Brook Valley, Canton Center, Connecticut. *Am. J. Sci.* 255:517–22.

Potter, P. E., and Pettijohn, F. J. 1963. *Paleocurrents and basin analysis.* Berlin: Springer-Verlag.

Pratt, L. M.; Vuletich, A. K.; and Burruss, R. C. 1986. Petroleum generation and migration in Lower Jurassic lacustrine sequences, Hartford basin, Connecticut and Massachusetts. In *U.S. Geological Survey research on energy resources, program and abstracts,* ed. L. M. H. Carter, p. 57–58. U.S. Geol. Survey Circular 974.

Ratcliffe, N. M. 1971. The Ramapo fault system in New York and adjacent New Jersey: a case of tectonic heredity. *Bull. Geol. Soc. Am.* 82:125–42.

Raymond, P. E. 1927*a*. The significance of red color in sediments. *Am. J. Sci.* 13:234–51.

————. 1927*b*. The significance of red color in sediments: discussion and communications. *Am. J. Sci.* 14:157–58.

Reading, H. G., ed. 1978. *Sedimentary environments and facies.* New York: Elsevier.

Redfield, J. H. 1836. Fossil fishes of Connecticut and Massachusetts, with a notice of an undescribed genus. *New York Lyc. Nat. Hist. Ann.* 4:35–40 [1848].

Redfield, W. C. 1856. On the relations of the fossil fishes of the sandstone of Connecticut, and other Atlantic states, to the Liassic and Jurassic periods. *Proc. Am. Ass'n Adv. Sci.* 10(2):180–88.

Reeside, J. B. et al. 1957. Correlation of the Triassic formations of North America, exclusive of Canada. *Bull. Geol. Soc. Am.* 68:1451–1514.

Reynolds, D. D., and Leavitt, D. H. 1927. A scree of Triassic age. *Am. J. Sci.* 13:167–71.

Rice, W. N., and Gregory, H. E. 1906. *Manual of the geology of Connecticut.* State Geol. and Nat. Hist. Survey of Conn. Bull. 6.

Roberts, J. K. 1928. *The geology of the Virginia Triassic.* Virginia Geol. Survey Bull. 29.

Robinson, P., and Luttrell, G. W. 1985. *Revision of some stratigraphic names in central Massachusetts.* U.S. Geol. Survey Bull. 1605-A, pp. A71–A78.

Rodgers, J. 1967. Chronology of tectonic movements in the Appalachian region of eastern North America. *Am. J. Sci.* 265:408-27.

———. 1985. *Bedrock geological map of Connecticut, 1:125,000.* State Geol. and Nat. Hist. Survey of Conn. and U..S. Geol. Survey.

Rodgers, J.; Gates, R. M.; and Rosenfeld, J. L. 1959. *Explanatory text for preliminary geological map of Connecticut, 1956.* State Geol. and Nat. Hist. Survey of Conn. Bull. No. 84.

Rogers, H. D. 1840. *Description of the geology of the state of New Jersey, being a final report.* Philadelphia: C. Sherman and Co.

———. 1856. [Quoted on the connection between fossil footprints and the theory of progressive development.] *Annual of Scientific Discovery for 1856,* p. 314-16.

Rogers, H. D., and Rogers, W. B. 1842. [Quoted as to the origin of the inclined strata, in a report on a scientific meeting.] *Am. J. Sci.* 43:170-73.

Rogers, H. D.; Vanuxem, L.; Taylor, R. C.; Emmons, E.; and Conrad, T. A. 1841. Report on the ornithichnites or foot marks of extinct birds, in the New Red Sandstone of Massachusetts and Connecticut, observed and described by Prof. Hitchcock, Amherst. *Am. J. Sci.* 41:165-68.

Russell, I. C. 1878. On the physical history of the Triassic formation in New Jersey and the Connecticut Valley. *Annals N.Y. Acad. Sci.* 1:220-54.

———. 1880. On the former extent of the Triassic formation of the Atlantic states. *Am. Naturalist* 14:703-12.

———. 1885. *Geological history of Lake Lahontan.* U.S. Geol. Survey Monograph 11.

———. 1889a. *Subaerial decay of rocks and origin of the red color of certain formations.* U.S. Geol. Survey Bull. 52.

———. 1889b. The Newark system. *Am. Geologist* 3:178-82.

———. 1891. Has *Newark* priority as a group name. *Am. Geologist* 7:238-41.

———. 1892. *The Newark system.* U.S. Geol. Survey Bull. 85.

———. 1895. The Newark system. *Science* 1:266-68.

Russell, W. L. 1922. The structural and stratigraphic relations of the great Triassic fault of southern Connecticut. *Am. J. Sci.* 4:483-97.

Sanders, J. E. 1960. Structural history of Triassic rocks of the Connecticut Valley belt and its regional implications. *Trans. N.Y. Acad. Sci.* 23:119-32.

———. 1962. Strike-slip displacement faults in Triassic rocks in New Jersey. *Science* 136:40-42.

———. 1963. Late Triassic history of northeastern United States. *Am. J. Sci.* 261:501-24.

———. 1965. Primary sedimentary structures formed by turbidity currents and related resedimentation mechanisms. In *Primary sedimentary structures and their hydrodynamic interpretation,* ed. G. V. Middleton, p. 192-219. Soc. Econ. Paleont. Mineral. Special Pub. 12.

———. 1968. Stratigraphy and primary sedimentary structures of fine-grained, well-bedded strata, inferred lake deposits, Upper Triassic, central and southern Connecticut. In *Late Paleozoic and Mesozoic continental sedimentation, northeastern North America,* ed. G. deVries Klein, pp. 265-305. Geol. Soc. Am. Special Paper 106.

———. 1970. *Stratigraphy and structure of the Triassic strata of the Gaillard graben, south-central Connecticut.* State Geol. and Nat. Hist. Survey of Conn. Guidebook No. 3.

———. 1974. *Guidebook to field trips in Rockland County.* New York: N.Y. Petrol. Expl. Soc.

Sanders, J. E.; Guidotti, C. V.; and Wilde, P. 1963. *Foxon fault and Gaillard graben in the Triassic of southern Connecticut.* State Geol. and Nat. Hist. Survey of Conn. Report of Investigations 2.

Schamel, S., and Hubbard, I. G. 1985. Thermal maturity of Newark Supergroup basins from vitrinite reflectance and clay mineralogy [abstract]. *Bull. Am. Ass'n Petrol. Geol.* 69:1447.

Shepard, C. U. 1837. *Report on the geological survey of Connecticut.* New Haven: B. L. Hamlen.

Silliman, B. 1810. Sketch of the mineralogy of the town of New Haven. *Memoirs Conn. Acad. Arts and Sci.* 1:83-96.

————. 1818. New localities of agate, chalcedony, chabasie, stilbite, analcime, titanium, prehnite &c. *Am. J. Sci.* 1:134-35.

————. 1820. Sketches of a tour in the counties of New Haven and Litchfield in Connecticut, with notices of the geology, mineralogy, and scenery. *Am. J. Sci.* 2:201-35.

————. 1821. Miscellaneous observations relating to geology, mineralogy, and some connected topics. *Am. J. Sci.* 3:221-26.

————. 1830. Igneous origin of some trap rocks. *Am. J. Sci.* 17:119-30.

————. 1833. Consistency of geology with sacred history. In R. Bakewell, *An introduction to geology*, p. 389-466 [supplement by the editor]. New Haven: Hezekia Howe.

Silliman, B., Jr. 1844. [Remarks on the dip of the strata in the Hartford basin, in Abstract of the proceedings of the fifth session of the Association of American Geologists and Naturalists.] *Am. J. Sci.* 47:107-8.

Skinner, B. J., and Rodgers, J. 1985. Geology of southern Connecticut, east-west transect. In *Guidebook for field trips in Connecticut and adjacent areas of New York and Rhode Island, 77th New England Intercollegiate Geological Conference*, ed. R. J. Tracy, p. 210-18. State Geol. and Nat. Hist. Survey of Conn. Guidebook 6.

Smoot, J. P., and Olsen, P. E. 1985. Massive mudstones in basin analysis and paleoclimatic interpretation of the Newark Supergroup. *Proceedings of the second U.S. Geological Survey workshop on the early Mesozoic basins of the eastern United States*, ed. G. R. Robinson and A. J. Froelich, p. 29-33. U.S. Geol. Survey Circular 946.

Smoot, J. P.; LeTourneau, P. M.; Turner-Peterson, C. M.; and Olsen, P. E. 1985. Sandstone and conglomerate shoreline deposits in Triassic-Jurassic Newark and Hartford basins of Newark Supergroup [abstract]. *Bull. Am. Ass'n Petrol. Geol.* 69:1448.

Sorby, H. C. 1859. On the structures produced by the currents present during the deposition of stratified rocks. *The Geologist* 2:137-47.

————. 1908. On the application of quantitative methods to the study of the structure and history of rocks. *Quart. J. Geol. Soc. London* 64:171-233.

Stevens, R. L., and Hubert, J. F. 1980. Alluvial fans, braided rivers, and lakes in a fault-bounded semiarid valley: Sugarloaf Arkose (Late Triassic-Early Jurassic), Newark Supergroup, Deerfield basin, Massachusetts. *Northeastern Geol.* 2:100-117.

Swanson, M. T. 1986. Pre-existing fault control for Mesozoic basin formation in eastern North America. *Geology* 14:419-22.

Teichert, C. 1958. Concepts of facies. *Bull. Am. Ass'n Petrol. Geol.* 42:2718-44.

Twain, M. 1896. *Life on the Mississippi.* Reprint 1972. New York: Bantam.

Van Houten, F. B. 1962. Cyclic sedimentation and the origin of analcime-rich Upper Triassic Lockatong Formation, west-central New Jersey and adjacent Pennsylvania. *Am. J. Sci.* 260:561-76.

————. 1964. Cyclic lacustrine sedimentation, Upper Triassic Lockatong Formation, central New Jersey and adjacent Pennsylvania. In *Symposium on cyclic sedimentation*, ed. D. F. Merriam, p. 497-531. Kansas Geol. Survey Bull. 169.

————. 1965. Composition of Triassic Lockatong and associated formations of Newark Group, central New Jersey and adjacent Pennsylvania. *Am. J. Sci.* 263:825-63.

————. 1968. Iron oxides in red beds. *Bull. Geol. Soc. Am.* 79:399-416.

————. 1973. Origin of red beds, a review—1961-1973. *Ann. Rev. of Earth and Planet. Sci.* 1:39-61.

————. 1977. Triassic-Liassic deposits of Morocco and eastern North America: comparison. *Bull. Am. Ass'n Petrol. Geol.* 61:79-99.

Visher, G. S. 1965. Use of vertical profile in environmental reconstruction. *Bull. Am. Ass'n Petrol. Geol.* 49:41-61.

Wadell, H. 1931. Sedimentation and sedimentology. *Science* 75:20.

————. 1933. Sedimentation and sedimentology. *Science* 77:536–37.

Walker, T. R. 1964. In situ formation of red beds in an arid to semi-arid climate [abstract]. In *Abstracts for 1963*, p. 174–75. Geol. Soc. Am. Sp. Paper 76.

————. 1967. Formation of red beds in modern and ancient deserts. *Bull. Geol. Soc. Am.* 78:353–68.

Ward, L. F. 1891. The plant-bearing beds of the American Trias. *Bull. Geol. Soc. Am.* 3:23–31.

Watts, A. B. 1981. The U.S. Atlantic continental margin: subsidence history, crustal structure, and thermal evolution. In *Geology of passive continental margins, history, structure, and sedimentologic record*, ed. A. W. Bally et al., p. 2-1-2-75. Am. Ass'n Petrol. Geol. Short Course Education Note Series 19.

Weddle, T. K., and Hubert, J. F. 1983. Petrology of Upper Triassic sandstones of the Newark Supergroup in the northern Newark, Pomperaug, Hartford, and Deerfield basins. *Northeastern Geol.* 5:8–22.

Wells, D. A. 1852. On the origin of stratification. *Proc. Am. Ass'n Adv. Sci.* 6:297–99.

Wessel, J. M. 1969. *Sedimentary history of Upper Triassic alluvial fan complexes in north-central Massachusetts.* Dept. of Geol: and Geog., Univ. of Mass., Contribution 2.

Wessel, J. M.; Hand, B. M.; and Hayes, M. O. 1967. Sedimentary features of the Triassic rocks in northern Massachusetts. In *Guidebook for field trips in the Connecticut Valley of Massachusetts: 59th New England Intercollegiate Geological Conference*, ed. P. Robinson, p. 154–65.

Wheeler, G. 1937. *The west wall of the New England Triassic lowland.* State Geol. and Nat. Hist. Survey of Conn. Bull. 58.

Willard, M. E. 1951. *Bedrock geology of the Mount Toby quadrangle, Massachusetts.* U.S. Geol. Survey Map GQ8.

————. 1952. *Bedrock geology of the Greenfield quadrangle, Massachusetts.* U.S. Geol. Survey Map GQ10.

Wise, D. U. 1982. New fault and fracture domains of southwestern New England—hints on localization of the Mesozoic basins. In *Geotechnology in Massachusetts: Proceedings Univ. of Mass. Conference March 1980*, ed. O. C. Farquhar, p. 447–53.

Zen, E-an, ed. 1983. *Bedrock geology map of Massachusetts, 1:250,000.* U.S. Geol. Survey and Mass. Dept. Public Works.

References, Part II

Amade, E. 1965. Le gisement de potasse triasique de Khémisset. *Maroc, Mines et Géologie, Rabat* 23:23–48.

April, R. H. 1981. Clay petrology of the Upper Triassic/Lower Jurassic terrestrial strata of the Newark Supergroup, Connecticut Valley, USA. *Sedimentary Geol.* 29:283–307.

Arguden, A. T., and Rodolfo, K. S. 1986. Sedimentary facies and tectonic implications of lower Mesozoic alluvial-fan conglomerates of the Newark basin, northeastern United States. *Sedimentary Geology* 51:97–118.

Arthurton, R. S. 1973. Experimentally produced halite compared with Triassic layered halite-rock from Cheshire, England. *Sedimentology* 20:145–60.

———. 1980. Rhythmic sedimentary sequences in the Triassic Keuper Marl (Mercia Mudstone Group) of Cheshire, northwest England. *Geol. J.* 15:43–58.

Audley-Charles, M. G. 1970. Triassic palaeogeography of the British Isles. *Quart. J. Geol. Soc. London* 126:49–89.

Bain, G. L., and Harvey, B. W. 1977. *Field guide to the geology of the Durham Triassic basin*. 40th Anniversary Meeting, Carolina Geological Society.

Bain, G. W. 1941. The Holyoke range and Connecticut valley structures. *Am. J. Sci.* 239:261–75.

Barrell, J. 1915. *Central Connecticut in the geologic past*. State Geol. and Nat. Hist. Survey of Conn. Bull. 23.

Beauchamp, J. 1983. Le Permien et le Trias marocains. In *Colloque sur le Permien et le Trias du Maroc*, pp. 1–21. Bulletin de la Faculté des Sciences, Université Cadi Ayyad, Marrakech, Section des Sciences de la Terre, Numéro Spécial 1.

Bhattacharyya, D. P., and Lorenz, J. C. 1983. Different depositional settings of the Nubian litho-facies in Libya and Egypt. In *Modern and ancient fluvial systems*, ed. J. D. Collinson and J. Lewin, p. 435–48. Int'l Ass'n Sedimentologists Special Pub. 6.

Birkelund, T., and Perch-Nielsen, K. 1976. Late Paleozoic–Mesozoic evolution of central East Greenland. In *The geology of Greenland*, ed. A. Escher and W. S. Watts, p. 304–39. Geol. Survey of Greenland.

Biron, P. E. 1982. Le Permo-Trias de la région de l'Ourika (Haut-Atlas de Marrakech, Maroc). Unpub. thesis from l'Université Scientifique et Médical de Grenoble.

Blissenbach, E. 1954. Geology of alluvial fans in semi-arid regions. *Bull. Geol. Soc. Am.* 65:175–90.

Blodgett, R. H. 1985. Paleovertisols—their utility in reconstructing ancient fluvial floodplain sequences [abstract]. In *Third International Fluvial Sedimentology Conference*, abstracts, p. 10.

Bluck, B. J. 1965. The sedimentary history of some Triassic conglomerates in the vale of Glamorgan, South Wales. *Sedimentology* 4:225–45.

———. 1967. Deposition of some upper Old Red Sandstone conglomerates in the Clyde area: a study in the significance of bedding. *Scottish J. Geol.* 3:139–67.

Bosellini, A., and Hsü, K. J. 1973. Mediterranean plate tectonics and Triassic palaeogeography. *Nature* 244:144–46.

Boyer, P. S. 1979. Trace fossils *Biformites* and *Fustiglyphus* from the Jurassic of New Jersey. *Bull. N.J. Acad. Sci.* 24:73–77.

Bromley, R. G., and Asgaard, U. 1972. Notes on Greenland trace fossils. *Rapp. Grønlands Geol. Unders.* 49:7–13.

———. 1979. Triassic freshwater ichnocoenoses from Carlsberg Fjord, East Greenland. *Palaeogeogr., Palaeoclim., Palaeoecol.* 28:39–80.

Brown, R. H. 1980. Triassic rocks of Argana Valley, southern Morocco, and their regional structural implications. *Bull. Am. Ass'n Petrol. Geol.* 64:988–1003.

Bull, W. B. 1972. Recognition of alluvial-fan deposits in the stratigraphical record. In *Recognition of ancient sedimentary environments*, ed. J. K. Rigby and W. Hamblin, p. 63–83. Soc. Econ. Paleont. Mineral. Special Pub. 16.

Carozzi, A. V. 1964. Complex ooids from Triassic lake deposits, Virginia. *Am. J. Sci.* 262:231–41.

Chadwick, R. A. 1985. Permian, Mesozoic, and Cenozoic structural evolution of England and Wales in relation to the principles of extension and inversion tectonics. In *Atlas of onshore sedimentary basins in England and Wales*, ed. A. Whittaker, p. 9–25. Glasgow: British Geological Survey.

Clemmensen, L. B. 1976. Tidally influenced deltaic sequences from the Kap Stewart Formation (Rhaetic–Liassic), Scoresby Land, East Greenland. *Bull. Geol. Soc. Denmark* 25:1–13.

———. 1977. Stratigraphical and sedimentological studies of Triassic rocks in central East Greenland. *Rapp. Grønlands Geol. Unders.* 85:89–97.

———. 1978a. Alternating aeolian, sabkha, and shallow-lake deposits from the Middle Triassic Gipsdalen Formation, Scoresby Land, East Greenland. *Palaeogeogr., Palaeoclim., Palaeoecol.* 24:111–35.

———. 1978b. Lacustrine facies and stromatolites from the Middle Triassic of East Greenland. *J. Sed. Petrol.* 48:1111–28.

———. 1979. Triassic lacustrine red-beds and palaeoclimate: the "Buntsandstein" of Helgoland and the Malmros Klint Member of East Greenland. *Geologische Rundschau* 68:748–74.

———. 1980. *Triassic rift sedimentation and paleogeography of central East Greenland.* Grønlands Geol. Unders. Bull. 138.

Clemmensen, L. B.; Holser, W. T.; and Winter, D. 1984. Stable isotope study through Permian–Triassic boundary in East Greenland. *Bull. Geol. Soc. Denmark* 33:253–60.

Cogney, G., and Faugères, J. C. 1975. Précisions sur la mise en place des épanchements basaltiques des formations Triasiques de la bordure septentrionale du Maroc central. *Bull. Soc. Géol. France* 17:721–33.

Cogney, G.; Normand, M.; Termier, H.; and Termier, G. 1974. Observations sur le basalte du bassin triasique de Rommani-Maaziz (Maroc occidental). *Notes Service Géol. Maroc* 36:153–73.

Cogney, G.; Termier, H.; and Termier, G. 1971. Sur la présence de "pillow-lavas" dans la basalte du Permo-Trias au Maroc central. *C. R. Acad. Sci. Paris* 273:446–49.

Collinson, J. D. 1978. Lakes. In *Sedimentary environments and facies*, ed. H. A. Reading, p. 61–79. New York: Elsevier.

Cook, R. U., and Warren, A. 1973. *Geomorphology in deserts.* London: Batsford.

Cornet, B., and Traverse, A. 1975. Palynological contributions to the chronology and stratigraphy of the Hartford basin in Connecticut and Massachusetts. *Geoscience and Man* 11:1–33.

Cornet, B.; Traverse, A.; and McDonald, N. G. 1973. Fossil spores, pollen, and fishes from Connecticut indicate Early Jurassic age for part of the Newark Group. *Science* 182:1243–46.

Craig, G. Y. 1965. Permian and Triassic. In *The geology of Scotland,* ed. G. Y. Craig, p. 383–400. Edinburgh: Oliver and Boyd.

Demicco, R. V., and Gierlowski-Kordesch, E. G. 1986. Facies sequences of a semi-arid closed basin: the Lower Jurassic East Berlin Formation of the Hartford basin, New England, USA. *Sedimentology* 33:107–18.

Dumbleton, M. J., and West, G. 1967. *Studies of the Keuper Marl: stability of aggregation under weathering.* Road Research Laboratory, Ministry of Transport, RRL Report LR85.

Emerson, B. K. 1917. *Geology of Massachusetts and Rhode Island.* U.S. Geol. Survey Bull. 597.

Evans, W. B. 1970. The Triassic salt deposits of north-western England. *Quart. J. Geol. Soc. London* 126:103–23.

Evans, W. B.; Wilson, A. A.; Taylor, B. J.; and Price, D. 1968. *Geology of the country around Macclesfield, Congleton, Crewe, and Middlewich.* Memoirs of the Geol. Survey of Great Britain.

Fairbridge, R. W. 1979. Vertical crustal movements and the rifting of continents. *Geologie en Mijnbouw* 58:273–76.

Fitch, F. J.; Miller, J. A.; and Thompson, D. B. 1966. The paleogeographic significance of isotopic age determinations on detrital micas from the Triassic of the Stockport–Macclesfield district, Cheshire, England. *Palaeogeogr., Palaeoclim., Palaeoecol.* 2:281–312.

Flores, R. M., and Warwick, P. D. 1984. Dynamics of coal deposition in intermontane alluvial paleoenvironments, Eocene Wasatch Formation, Powder River basin, Wyoming. In *1984 Symposium on the geology of Rocky Mountain coal,* p. 184–99. N. Dakota Geol. Soc. Pub. 84-1.

Gierlowski-Kordesch, E. G. 1985. Sedimentology and trace fossil paleoecology of the Lower Jurassic East Berlin Formation, Hartford basin, Connecticut and Massachusetts. Ph.D. dissertation, Case Western Reserve University.

Gilchrist, J. M. 1979. Sedimentology of the Lower to Middle Jurassic Portland Arkose of central Connecticut. Master's thesis, University of Massachusetts.

Glaeser, J. D. 1966. *Provenance, dispersal, and depositional environments of Triassic sediments in the Newark–Gettysburg basin.* Penn. Geol. Survey Bull. G 43.

Glennie, K. W. 1984. The structural framework and the pre-Permian history of the North Sea area. In *Introduction to the petroleum geology of the North Sea,* ed. K. W. Glennie, p. 17–39. London: Blackwell Scientific Publications.

Goodwin, B. K.; Weems, R. E.; Wilkes, G. P.; Froelich, A. J.; and Smoot, J. P. 1985. *Guidebook to the geology of the Richmond and Taylorsville basins, east-central Virginia.* Eastern Section Am. Ass'n Petrol. Geol. Mtg. Fieldtrip 4.

Gore, P. J. W. 1985a. Triassic floodplain lake in the Chatham Group (Newark Supergroup) of the Durham sub-basin of the Deep River basin, N.C.: comparison to the Culpeper basin, Va. [abstract]. *Geol. Soc. Am. Abstracts with Programs* 17:94.

———. 1985b. Lacustrine sequences of Triassic-Jurassic age in the Culpeper basin, Virginia, and the Deep River basin, N.C. [abstract]. *Geol. Soc. Am. Abstracts with Programs* 17:21.

———. 1986. Early diagenetic nodules, compaction, and secondary lamination in Early Jurassic lacustrine black shale, Culpeper basin, Virginia [abstract]. *Bull. Am. Ass'n Petrol. Geol.* 70:596.

Gould, S. J. 1986. Evolution and the triumph of homology, or why history matters. *Am. Scientist* 74:60–69.

Habicht, J. K. A. 1979. *Paleoclimate, paleomagnetism, and continental drift.* Am. Ass'n Petrol. Geol. Studies in Geology 9.

Hall, J. 1882. [President's address to the American Association for the Advancement of Science, written in 1857.] *Proc. Am. Ass'n Adv. Sci.* 31:29–71.

Hallam, A. 1984. Pre-Quaternary sea-level changes. *Ann. Rev. of Earth and Planet. Sci.* 12:205–43.

Hand, B. M.; Wessel, J. M.; and Hayes, M. O. 1969. Antidunes in the Mount Toby Conglomerate (Triassic), Massachusetts. *J. Sed. Petrol.* 39:1310–16.

Hardie, L. A. 1984. Evaporites: marine or non-marine? *Am. J. Sci.* 284:193–240.

Hardie, L. A.; Smoot, J. P.; and Eugster, H. P. 1978. Saline lakes and their deposits: a sedimentological approach. In *Modern and ancient lake sediments*, ed. A. Matter and M. E. Tucker, p. 7–41. Int'l Ass'n Sedimentologists Special Pub. 2.

Hardie, L. A.; Lowenstein, T. K.; and Spencer, R. J. 1983. The problem of distinguishing between primary and secondary features in evaporites. In *Sixth international symposium on salt,* ed. B. C. Schreiber and H. L. Harner, p. 11–39.

Harding, A. G., and Brown, R. H. 1975. Structural controls over facies distributions in a Late Triassic–Early Jurassic carbonate sulfate-redbed sequence in southwestern Morocco, and its relationship to the opening of the Atlantic [abstract]. *Geol. Soc. Am. Abstracts with Programs* 7:1099–1100.

Haslam, J.; Allberry, E. C.; and Moses, G. 1950. The bromine content of the Cheshire salt deposit and of some borehole and other brines. *The Analyst* 75:352–56.

Henson, M. R. 1970. The Triassic rocks of South Devon. *Proc. Ussher Soc.* 2:172–77.

Hentz, T. F. 1985. Early Jurassic sedimentation of a rift-valley lake: Culpeper basin, northern Virginia. *Bull. Geol. Soc. Am.* 96:92–107.

Hogg, S. E. 1982. Sheetfloods, sheetwash, sheetflow, or . . . ? *Earth-Science Reviews* 18:59–76.

Holloway, S. 1985. Triassic: Sherwood Sandstone Group [chapter 5], and Mercia Mudstone and Penarth Groups [chapter 6]. In *Atlas of onshore sedimentary basins in England and Wales,* ed. A. Whittaker, p. 31–36. Glasgow: British Geological Survey.

Holser, W. T., and Wilgus, C. K. 1981. Bromide profiles of the Röt salt, Triassic of northern Europe, as evidence of its marine origin. *N. Jahrbk. Miner. Mh.* 6:267–76.

Hubert, J. F. 1977. Paleosol caliche in the New Haven Arkose, Connecticut: record of semi-aridity in Late Triassic–Early Jurassic time. *Geology* 4:302–4.

———. 1978. Paleosol caliche in the New Haven Arkose, Newark Group, Connecticut. *Palaeogeogr., Palaeoclim., Palaeoecol.* 24:151–68.

———. 1985. Braided-river deposits of Late Triassic age, Wolfville redbeds, Nova Scotia, Canada [abstract]. In *Third International Fluvial Sedimentology Conference,* abstracts, p. 23.

Hubert, J. F., and Forlenza, M. F. 1987. Sedimentology of braided-river deposits in Upper Triassic Wolfville redbeds, southern shore of Cobequid Bay, Nova Scotia, Canada. In *Triassic–Jurassic rifting and the opening of the Atlantic Ocean,* ed. W. Manspeizer, in press. Amsterdam: Elsevier.

Hubert, J. F.; Forlenza, M. F.; Perkins, R. E.; and Mertz, K. A. 1983. Upper Triassic alluvial fans, braided rivers, and eolian dunes of the Wolfville redbeds, Fundy rift valley, Nova Scotia [abstract]. *Geol. Soc. Am. Abstracts with Programs* 15:121.

Hubert, J. F.; Gilchrist, J. M.; and Reed, A. A. 1982. Jurassic redbeds of the Connecticut Valley: (1) brownstones of the Portland Formation; and (2) playa–playa lake–oligomictic lake model for parts of the East Berlin, Shuttle Meadow, and Portland Formations. In *Guidebook for field trips in Connecticut and south-central Massachusetts,* ed. R. Joesten, and S. S. Quarrier, p. 103–41. State Geol. and Nat. Hist. Survey of Conn. Guidebook 5.

Hubert, J. F., and Hyde, M. G. 1982. Sheet-flow deposits of graded beds and mudstones on an alluvial sandflat-playa system: Upper Triassic Blomidon redbeds, St. Mary's Bay, Nova Scotia. *Sedimentology* 29:457–74.

Hubert, J. F., and Mertz, K. A. 1980. Eolian dune field of Late Triassic age, Fundy basin, Nova Scotia. *Geology* 8:516–19.

———. 1984. Eolian sandstones in Upper Triassic–Lower Jurassic redbeds of the Fundy basin, Nova Scotia. *J. Sed. Petrol.* 54:798–810.

Hubert, J. F., and Reed, A. A. 1978. Red-bed diagenesis in the East Berlin Formation, Newark Group, Connecticut Valley. *J. Sed. Petrol.* 48:175–84.

Hubert, J. F.; Reed, A. A.; and Carey, P. J. 1976. Paleogeography of the East Berlin Formation, Newark Group, Connecticut Valley. *Am. J. Sci.* 276:1183–1207.

Hubert, J. F.; Reed, A. A.; Dowdall, W. L.; and Gilchrist, J. M. 1978. *Guide to the redbeds of central Connecticut: 1978 field trip, eastern section of Soc. Econ. Paleont. Mineral.* Dept. of Geol. and Geog., Univ. of Mass., Contribution 32.

Ireland, R. J.; Pollard, J. F; Steel, R. J.; and Thompson, D. B. 1978. Intertidal sediments and trace fossils from the Waterstones (Scythian–Anisian?) at Daresbury, Cheshire. *Proc. Yorkshire Geol. Soc.* 41:399–432.

Jansa, L. F; Bujak, J. P.; and Williams, G. L. 1980. Upper Triassic salt deposits of the western North Atlantic. *Canadian J. Earth Sci.* 17:547–59.

Jeans, C. V. 1978. The origin of the Triassic clay assemblages of Europe, with special reference to the Keuper Marl and Rhaetic of parts of England. *Phil. Trans. Royal Soc. London* 289:549–639.

Jones, D. F. 1975. Stratigraphy, environments of deposition, petrology, age and provenance of the basal red beds of the Argana Valley, western High Atlas Mountains, Morocco. Master's thesis, New Mexico Institute of Mining and Technology.

Katz, S. B., and Smoot, J. P. 1985. Sedimentary features and soil-like fabrics in cycles of upper Lockatong Formation, Newark Supergroup (Upper Triassic), New Jersey and Pennsylvania [abstract]. *Am. Ass'n Petrol. Geol.* 69:271–72.

King, P. B. 1961. Systematic patterns of Triassic dikes in the Appalachian region. In *Short papers in the geologic and hydrologic sciences—Geological Survey research 1961*, p. 93–95. U.S. Geol. Survey Prof. Paper 424-B.

Klein, G. de V. 1962a. Sedimentary structures in the Keuper Marl (Upper Triassic). *Geol. Mag.* 99:137–44.

———. 1962b. Triassic sedimentation, maritime provinces, Canada. *Bull. Geol. Soc. Am.* 73:1127–46.

Klitgord, K. D., and Hutchinson, D. R. 1985. Distribution and geophysical signatures of early Mesozoic rift basins beneath the U.S. Atlantic continental margin. In *Proceedings of the second U.S. Geological Survey workshop on the early Mesozoic basins of the eastern United States*, ed. G. R. Robinson and A. J. Froelich, p. 45–61. U.S. Geol. Survey Circular 946.

Kocurek, G., and Dott, R. H. 1981. Distinctions and uses of stratification types in the interpretation of eolian sand. *J. Sed. Petrol.* 51:579–95.

Kreisa, R. D., and Moiola, R. J. 1986. Sigmoidal tidal bundles and other tide-generated sedimentary structures of the Curtis Formation, Utah. *Bull. Geol. Soc. Am.* 97:381–87.

Krynine, P. D. 1950. *Petrology, stratigraphy, and origin of the Triassic sedimentary rocks of Connecticut.* State Geol. and Nat. Hist. Survey of Conn. Bull. 73.

Krystinik, L. F. 1986. Personal communication.

Last, W. M. 1984. Sedimentology of playa lakes of the northern Great Plains. *Canadian J. Earth Sci.* 21:107–25.

Lee, K. Y. 1977. *Triassic stratigraphy in the northern part of the Culpeper basin, Virginia and Maryland.* U.S. Geol. Survey Bull. 1422-C.

Leith, C. J., and Custer, R. L. P. 1968. Triassic paleocurrents in the Durham basin, North Carolina [abstract]. In *Abstracts for 1967*, p. 484–85. Geol. Soc. Am. Special Paper 115.

LeTourneau, P. M. 1985a. The sedimentology and stratigraphy of the Lower Jurassic Portland Formation, central Connecticut. Master's thesis, Wesleyan University.

———. 1985b. Lithofacies analysis of the Portland Formation in central Connecticut [abstract]. *Geol. Soc. Am. Abstracts with Programs* 17:31.

———. 1985c. Alluvial fan development in the Lower Jurassic Portland Formation, central

Connecticut—implications for tectonics and climate. In *Proceedings of the second U.S. Geological Survey workshop on the early Mesozoic basins of the eastern United States,* ed. G. R. Robinson and A. J. Froelich, p. 17–26. U.S. Geol. Survey Circular 946.

LeTourneau, P. M., and McDonald, N. G. 1985. The sedimentology, stratigraphy, and paleontology of the Lower Jurassic Portland Formation, Hartford basin, central Connecticut. In *Guidebook for field trips in Connecticut and adjacent areas of New York and Rhode Island,* p. 353–91. ed. R. J. Tracy, State Geol. and Nat. Hist. Survey of Conn. Guidebook 6.

———. 1986. Sedimentology and paleontology of a fluvial–deltaic–lacustrine sequence, Portland Formation (Jurassic), Hartford basin, Connecticut [abstract]. *Geol. Soc. Am. Abstracts with Programs* 18:30.

LeTourneau, P. M., and Smoot, J. P. 1985. Comparison of ancient and modern lake-margin deposits from the Lower Jurassic Portland Formation, Connecticut, and Walker Lake, Nevada [abstract]. *Geol. Soc. Am. Abstracts with Programs* 17:31.

Lindholm, R. C. 1978a. Triassic–Jurassic faulting in eastern North America—a model based on pre-Triassic structure. *Geology* 6:365–68.

———. 1978b. Tectonic control of sedimentation in Triassic–Jurassic Culpeper basin, Virginia [abstract]. *Bull. Am. Ass'n Petrol. Geol.* 62:537.

———. 1979. Geologic history and stratigraphy of the Triassic–Jurassic Culpeper basin, Virginia. *Bull. Geol. Soc. Am.* 90:1702–36.

Lindholm, R. C.; Hazlett, J. M.; and Fagin, S. W. 1979. Petrology of Triassic–Jurassic conglomerates in the Culpeper basin, Virginia. *J. Sed. Petrol.* 41:1245–62.

Lorenz, J. C. 1976. Triassic sediments and basin structure of the Kerrouchen basin, central Morocco. *J. Sed. Petrol.* 46:897–905.

Lowenstein, T. K., and Hardie, L. A. 1985. Criteria for the recognition of salt-pan evaporites. *Sedimentology* 32:627–44.

Lucas, J. 1962. *La transformation des minéraux argileaux dans la sédimentation; études sur les argiles du Trias.* Mém. Serv. Carte Géol. Alsace-Lorraine, Strasbourg 23.

McCabe, P. J. 1984. Depositional environments of coal and coal-bearing strata. In *Sedimentology of coal and coal-bearing sequences,* ed. R. A. Rahmani and R. M. Flores, pp. 13–42. Int'l Ass'n Sedimentologists Special Pub. 7.

McDonald, N. G. 1985. New discoveries of Jurassic invertebrates in the Connecticut Valley: implications for lacustrine paleoecology [abstract]. *Geol. Soc. Am. Abstracts with Programs* 17:53.

McKee, E. D. 1966. Dune structures. *Sedimentology* 7:1–69.

McKee, E. D., ed. 1979. *A study of global sand seas.* U.S. Geol. Survey Prof. Paper 1052.

McKee, E. D.; Crosby, E. J.; and Berryhill, H. L., Jr. 1967. Flood deposits, Bijou Creek, Colorado, June, 1965. *J. Sed. Petrol.* 37:829–51.

Maclure, W. 1809. Observations on the geology of the United States, explanatory of a geological map. *Trans. Am. Phil. Soc.* 6(2):411–28. Republished (1817) in monograph form. Philadelphia: A. Small.

Manspeizer, W. 1980. Rift tectonics inferred from volcanic and clastic structures. In *Field studies of New Jersey geology and guide to field trips: 52nd annual meeting of the New York State Geological Association,* ed. W. Manspeizer, p. 314–350.

———. 1981. Early Mesozoic basins of central Atlantic passive margins. In *Geology of passive continental margins: history, structure, and sedimentologic record,* ed. A. W. Bally, p. 1–60. Am. Ass'n Petrol. Geol. Short Course Education Notes Series 19.

———. 1982. Triassic–Liassic basins and climate of the Atlantic passive margins. *Geologische Rundschau* 71:895–917.

Manspeizer, W.; Puffer, J. H.; and Couzminer, H. L. 1978. Separation of Morocco and eastern North America: a Triassic–Liassic stratigraphic record. *Bull. Geol. Soc. Am.* 89:901–20.

Martini, I. P.; Rau, A.; and Tongiorgi, M. 1986. Syntectonic sedimentation in a Middle Triassic rift, northern Apennines, Italy. *Sedimentary Geol.* 47:191-219.

Mattis, A. F. 1974. Upper Triassic lacustrine carbonates, central High Atlas Mountains, Morocco [abstract]. *Geol. Soc. Am. Abstracts with Programs* 6:53.

———. 1977. Nonmarine Triassic sedimentation, central High Atlas Mountains, Morocco. *J. Sed. Petrology* 47:107-19.

May, P. R. 1971. Pattern of Triassic-Jurassic diabase dikes around the North Atlantic in the context of predrift position of the continents. *Bull. Geol. Soc. Am.* 82:1285-92.

Miall, A. D. 1978a. Fluvial sedimentology: an historical review. In *Fluvial sedimentology*, ed. A. D. Miall, p. 1-47. Canadian Soc. Petrol. Geol. Memoir 5.

———. 1978b. Lithofacies types and vertical profile models in braided river models: a summary. In *Fluvial sedimentology*, ed. A. D. Miall, p. 597-604. Canadian Soc. Petrol. Geol. Memoir 5.

———. 1985. Architectural-element analysis: a new method of facies analysis applied to fluvial deposits. *Earth-Science Reviews* 22:261-308.

Nadon, G. C., and Middleton, G. V. 1985. The stratigraphy and sedimentology of the Fundy Group (Triassic) of the St. Martins area, New Brunswick. *Canadian J. Earth Sci.* 22:1183-1203.

Nilsen, T. H., ed. 1985. *Modern and ancient alluvial fan deposits.* New York: Van Nostrand Reinhold.

Olsen, P. E. 1977. Triangle brick quarry. In *Field guide to the geology of the Durham Triassic basin: Carolina Geological Society 40th anniversary meeting*, ed. G. L. Bain and B. W. Harvey, p. 59-60.

———. 1980. Fossil great lakes of the Newark Supergroup in New Jersey. *Field studies of New Jersey geology and guide to field trips: 52nd annual meeting of the New York State Geological Association*, ed. W. Manspeizer, p. 352-98.

———. 1985a. Significance of great lateral extent of thin units in Newark Supergroup (Lower Mesozoic, eastern North America) [abstract]. *Bull. Am. Ass'n Petrol. Geol.* 69:1444.

———. 1985b. Distribution of organic matter-rich lacustrine rocks in the early Mesozoic Newark Supergroup. In *Proceedings of the second U.S. Geological Survey workshop on the early Mesozoic basins of the eastern United States*, ed. G. R. Robinson and A. J. Froelich, pp. 61-64. U.S. Geol. Survey Circular 946.

———. 1986. A 40-million year lake record of early Mesozoic orbital climatic forcing. *Science* 243:842-48.

Parnell, J. 1983. Skeletal halites from the Jurassic of Massachusetts, and their significance. *Sedimentology* 30:711-15.

Pattison, J.; Smith, D. B.; and Warrington, G. 1973. A review of Late Permian and Early Triassic biostratigraphy in the British Isles. In *The Permian and Triassic systems and their mutual boundary*, ed. A. Logan and L. V. Hills, p. 220-60. Canadian Soc. Petrol. Geol. Memoir 2.

Petit, J.-P., and Beauchamp, J. 1986. Synsedimentary faulting and paleocurrent patterns in the Triassic sandstones of the High Atlas (Morocco). *Sedimentology* 33:817-29.

Picard, M. D., and High, L. R., Jr. 1972. Criteria for recognizing lacustrine rocks. In *Recognition of ancient sedimentary environments*, ed. J. K. Rigby and W. K. Hamblin, p. 108-45. Soc. Econ. Mineral. Special Pub. 16.

Porter, K. G., and Robbins, E. I. 1981. Zooplankton fecal pellets link fossil fuel and phosphate deposits. *Science* 212:931-33.

Pratt, L. M.; Vuletich, A. K.; and Burruss, R. C. 1986. Petroleum generation and migration in Lower Jurassic lacustrine sequences, Hartford basin, Connecticut and Massachusetts. In *U.S. Geological Survey research on energy resources, program and abstracts*, ed. L. M. H. Carter, p. 57-58. U.S. Geol. Survey Circular 946.

Prouty, W. F. 1931. Triassic deposits of the Durham basin, and their relation to other Triassic areas of eastern United States. *Am. J. Sci.* 21:473-90.

Ramos, A., and Sopeña, A. 1983. Gravel bars in low-sinuosity streams (Permian and Triassic, central Spain). In *Modern and ancient fluvial systems*, ed. J. D. Collinson and J. Lewin, International Association of Sedimentologists Special Publication 6:301–12.

Ramos, A., Sopeña, A., and Perez-Arlucea, M. 1986. Evolution of Bundtsandstein fluvial sedimentation in northwest Iberian ranges (central Spain). *Journal of Sedimentary Petrology* 56:862–75.

Randazzo, A. F., and Copeland, R. E. 1976. The geology of the northern portion of the Wadesboro Triassic basin, North Carolina. *Southeastern Geol.* 17:115–38.

Randazzo, A. F.; Swe, W.; and Wheeler, W. H. 1970. A study of tectonic influence on Triassic sedimentation—the Wadesboro basin, central piedmont. *J. Sed. Petrol.* 40:998–1006.

Ratcliffe, N. M. 1971. The Ramapo fault system in New York and adjacent New Jersey: a case of tectonic heredity. *Bull. Geol. Soc. Am.* 82:125–42.

Ratcliffe, N. M., and Burton, W. C. 1985. Fault reactivation models for origin of the Newark basin and studies related to eastern U.S. seismicity. In *Proceedings of the second U.S. Geological Survey workshop on the early Mesozoic basins of the eastern United States*, ed. G. R. Robinson and A. J. Froelich, p. 36–45. U.S. Geol. Survey Circular 946.

Ratcliffe, N. M., Burton, W. C., D'Angelo, R. M., and Costain, J. K. 1986. Low-angle extension faulting, reactivated mylonites, and seismic reflection geometry of the Newark basin margin in eastern Pennsylvania. *Geology* 14:766–70.

Reinemund, J. A. 1955. *Geology of the Deep River coal field, North Carolina.* U.S. Geol. Survey Prof. Paper 246.

Ressetar, R., and Taylor, K. 1987. Depositional history of the Richmond and Taylorsville basins, eastern Virginia. In *Triassic-Jurassic rifting and the opening of the Atlantic Ocean*, ed. W. Manspeizer, in press. Amsterdam: Elsevier.

Reynolds, D. D., and Leavitt, D. H. 1927. A scree of Triassic age. *Am. J. Sci.* 13:167–71.

Robbins, E. I. 1982. "Fossil Lake Danville": the paleoecology of a Late Triassic ecosystem on the North Carolina–Virginia border. Ph.D. thesis, Dept. Geosciences, Penn. State Univ., 400 p.

———. 1983. Accumulation of fossil fuels and metallic minerals in active and ancient rift lakes. *Tectonophysics* 94:633–58.

———. 1985a. Palynostratigraphy of coal-bearing sequences in early Mesozoic basins of the eastern United States. In *Proceedings of the second U.S. Geological Survey workshop on the early Mesozoic basins of the eastern United States*, ed. G. R. Robinson and A. J. Froelich, p. 27–29. U.S. Geol. Survey Circular 946.

———. 1985b. Personal communication.

Robbins, E. I.; Wilkes, G. P.; and Textoris, D. A. 1987. Coals of the Newark rift system. In *Triassic-Jurassic rifting and the opening of the Atlantic Ocean*, ed. W. Manspeizer, in press. Amsterdam: Elsevier.

Roberts, J. K. 1928. *The geology of the Virginia Triassic.* Virginia Geol. Survey Bull. 29.

Robinson, P. A. 1973. Palaeoclimatology and continental drift. In *Implications of continental drift to the earth sciences*, ed. D. H. Tarling and S. K. Runcorn, p. 451–76. London: Academic Press.

Rodgers, J. 1985. *Bedrock geological map of Connecticut, 1:125,000.* State Geol. and Nat. Hist. Survey of Conn. and U.S. Geol. Survey.

Rose, G. N., and Kent, P. E. 1955. A *Lingula*-bed in the Keuper of Nottinghamshire. *Geol. Mag.* 92:476–80.

Salvan, H. M. 1968a. L'évolution du problème des évaporites et ses conséquences sur l'interprétation des gisements marocains. *Mines et Géologie*, Rabat, 27:5–30.

———. 1968b. Les niveaux salifères marocains, leurs caractéristiques et leurs problèmes. In *Geology of saline deposits*, pp. 147–59. UNESCO Earth Sci. Series, No. 7.

———. 1974a. Les séries salifères du Trias marocain; caractères généraux et possibilités d'interprétation. *Bull. Soc. Géol. France* 7:724–31.

————. 1974*b*. Les séries salifères triasiques du Maroc; comparison avec les séries homologues d'Algèrie et de Tunisie: nouvelles possibilités d'interprétation. *Notes Service Géol. Maroc* 35:7–25.

————. 1983. Les évaporites triasiques du Maroc, caractères généraux-répartition-interprétation. In *Colloque sur le Permien et le Trias du Maroc*, pp. 73–84. Bulletin de la Faculté des Sciences, Université Cadi Ayyad, Marrakech, Section des Sciences de la Terre, Numéro Spécial 1.

Sanders, J. E. 1968. Stratigraphy and primary sedimentary structures of fine-grained, well-bedded strata, inferred lake deposits, Upper Triassic, central and southern Connecticut. In *Late Paleozoic and Mesozoic continental sedimentation, northeastern North America*, ed. G. de Vries Klein, p. 265–305. Geol. Soc. Am. Special Paper 106.

Schlee, J. S., and Jansa, L. F. 1981. The paleoenvironment and development of the eastern North American continental margin. In *Oceanologica Acta, Colloque C3, Géologie des marges continentales.* Supplement to vol. 4, p. 71–80.

Schutz, D. F. 1956. The geology of Pomperaug Valley, Connecticut. Senior thesis, Yale University.

Sitian, L.; Baofang, L.; Shigong, Y.; Jiafu, H.; and Zhen, L. 1984. Sedimentation and tectonic evolution of late Mesozoic faulted coal basins in north-eastern China. In *Sedimentology of coal and coal-bearing sequences*, ed. R. A. Rahmani and R. M. Flores, p. 387–406. Int'l Ass'n Sedimentologists Special Pub. 7.

Smoot, J. P. 1985. The closed-basin hypothesis of the Newark Supergroup. In *Proceedings of the second U.S. Geological Survey workshop on the early Mesozoic basins of the eastern United States*, ed. G. R. Robinson and A. J. Froelich, p. 4–10. U.S. Geol. Survey Circular 946.

Smoot, J. P.; LeTourneau, P. M.; Turner-Peterson, C. M.; and Olsen, P. E. 1985. Sandstone and conglomerate shoreline deposits in Triassic-Jurassic Newark and Hartford basins of Newark Supergroup [abstract]. *Bull. Am. Ass'n Petrol. Geol.* 69:1448.

Smoot, J. P., and Olsen, P. E. 1985. Massive mudstones in basin analysis and paleoclimatic interpretation of the Newark Supergroup. In *Proceedings of the second U.S. Geological Survey workshop on the early Mesozoic basins of the eastern United States*, ed. G. R. Robinson and A. J. Froelich, p. 29–33. U.S. Geol. Survey Circular 946.

Steel, R. J. 1974. New Red Sandstone floodplain and piedmont sedimentation in the Hebridean province, Scotland. *J. Sed. Petrol.* 44:336–57.

Steel, R. J., and Thompson, D. B. 1983. Structures and textures in Triassic braided stream conglomerates ("Bunter" Pebble Beds) in the Sherwood Sandstone Group, north Staffordshire, England. *Sedimentology* 30:341–67.

Steel, R. J., and Wilson, A. C. 1975. Sedimentation and tectonism (?Permo-Triassic) on the margin of the North Minch basin, Lewis. *J. Geol. Soc. London* 131:183–202.

Stévaux, J. 1971. Les faciès du Keuper en Aquitaine: Paléogéographie et dépendances avec leur substratum. *Bull. Centre. Rech. Pau* 5:357–61.

Stévaux, J., and Winnock, E. 1974. Les bassins du Trias et du Lias inférieur d'Aquitaine et leurs épisodes évaporitiques. *Bull. Géol. Soc. France* 16:679–95.

Stevens, R. L., and Hubert, J. F. 1980. Alluvial fans, braided rivers, and lakes in a fault-bounded semiarid valley: Sugarloaf Arkose (Late Triassic-Early Jurassic), Newark Supergroup, Deerfield basin, Massachusetts. *Northeastern Geol.* 2:100–17.

Swanson, M. T. 1982. Preliminary model for an early transform history in central Atlantic rifting. *Geology* 10:317–20.

————. 1986. Preexisting fault control for Mesozoic basin formation in eastern North America. *Geology* 14:419–22.

Taylor, S. R. 1983. A stable isotope study of the Mercia Mudstones (Keuper Marl) and associated sulphate horizons in the English Midlands. *Sedimentology* 30:11–31.

Thayer, P. A. 1970. Stratigraphy and geology of Dan River Triassic basin, North Carolina. *Southeastern Geol.* 12:1–31.

Thompson, D. B. 1969. Dome-shaped aeolian dunes in the Frodsham Member of the so-called "Keuper" Sandstone Formation (Scythian?-Anisian: Triassic) at Frodsham, Cheshire (England). *Sedimentary Geol.* 3:263–89.

———. 1970*a*. Sedimentation of the Triassic (Scythian) red pebbly sandstones in the Cheshire basin and its margins. *Geol. J.* 7:183–261.

———. 1970*b*. The stratigraphy of the so-called Keuper Sandstone Formation (Scythian?-Anisian) in the Permo–Triassic Cheshire basin. *Quart. J. Geol. Soc. London* 126:151–81.

Tucker, M. E. 1975. Vadose diagenetic fabrics in Triassic lacustrine limestones from South Wales. In *Extraits des publications du 9 Congress International de Sedimentologie, Nice*, vol. 7, p. 217–23.

———. 1977. The marginal Triassic deposits of South Wales: continental facies and paleogeography. *Geol. J.*, 12:169–88.

———. 1978. Triassic lacustrine sediments from South Wales: shore-zone clastics, evaporites, and carbonates. In *Modern and ancient lake sediments*, ed. A. Matter and M. E. Tucker, p. 205–24. Int'l Ass'n Sedimentologists Special Pub. 2.

———. 1981. Giant polygons in the Triassic salt of Cheshire, England: a thermal contraction model for their origin. *J. Sed. Petrol.* 51:779–86.

Tucker, M. E., and Burchette, T. P. 1977. Triassic dinosaur footprints from South Wales: their context and preservation. *Palaeogeogr., Palaeoclim., Palaeoecol.* 22:195–208.

Tucker, R. M., and Tucker, M. E. 1981. Evidence of synsedimentary tectonic movements in the Triassic halite of Cheshire. *Nature* 290:495–96.

Turner-Peterson, C. M. 1980. Sedimentology and uranium mineralization in the Triassic–Jurassic Newark basin, Pennsylvania and New Jersey. In *Uranium in sedimentary rocks: application of the facies concept to exploration*, ed. C. M. Turner-Peterson, p. 149–75. Soc. Econ. Paleont. Mineral.

Turner-Peterson, C. M., and Smoot, J. P. 1985. New thoughts on facies relationships in the Triassic Stockton and Lockatong Formations, Pennsylvania and New Jersey. In *Proceedings of the second U.S. Geological Survey workshop on the early Mesozoic basins of the eastern United States*, ed. G. R. Robinson and A. J. Froelich, p. 10–17. U.S. Geol. Survey Circular 946.

Vail, P. R.; Mitchum, R. M., Jr.; and Thompson, S., III. 1977. Seismic stratigraphy and global changes of sea level; 4: Global cycles of relative changes of sea level. In *Stratigraphic interpretation of seismic data*, ed. C. E. Payton, p. 83–97. Am. Ass'n Petrol. Geol. Memoir 26.

Van Houten, F. B. 1962. Cyclic sedimentation and the origin of analcime-rich Upper Triassic Lockatong Formation, west-central New Jersey and adjacent Pennsylvania. *Am. J. Sci.* 260:516–76.

———. 1964. Cyclic lacustrine sedimentation, Upper Triassic Lockatong Formation, central New Jersey and adjacent Pennsylvania. In *Symposium on cyclic sedimentation*, ed. D. F. Merriam, p. 497–531. Kansas Geol. Survey Bull. 169.

———. 1965*a*. Crystal casts in Upper Triassic Lockatong and Brunswick Formations. *Sedimentology* 4:301–13.

———. 1965*b*. Composition of Triassic Lockatong and associated formations of Newark Group, central New Jersey and adjacent Pennsylvania. *Am. J. Sci.* 263:825–63.

———. 1969. Late Triassic Newark Group, north central New Jersey and adjacent Pennsylvania and New York. In *Geology of selected areas in New Jersey and eastern Pennsylvania and guidebook of excursions*, ed. S. S. Subitsky, p. 314–47. Rutgers, N.J.: Rutgers Univ. Press.

———. 1977. Triassic–Liassic deposits of Morocco and eastern North America: comparison. *Bull. Am. Ass'n Petrol. Geol.* 61:79–99.

———. 1980. Late Triassic part of Newark Supergroup, Delaware River section, west-central

New Jersey. In *Field studies of New Jersey geology and guide to field trips: 52nd annual meeting of the New York State Geological Association,* ed. W. Manspeizer, p. 264–76.

Vetter, M., and Brakenridge, G. R. 1986. Hartford and Deerfield basin framework mineralogies: independent evidence for provinance, current directions, and tectonic history [abstract]. *Bull. Am. Ass'n Petrol. Geol.* 70:659.

Vogel, T. A.; Williams, E. R.; Preston, J. K.; and Walker, B. M. 1976. Origin of late Paleozoic plutonic massifs in Morocco. *Bull. Geol. Soc. Am.* 87:1753–62.

Warren, J. K., and Kendall, C. G. 1985. Comparison of sequences formed in marine sabkha (subaerial) and salina (subaqueous) settings—modern and ancient. *Bull. Am. Ass'n Petrol. Geol.* 69:1013–23.

Warrington, G. 1967. Correlation of the Keuper series of the Triassic by miospores. *Nature* 214:1323–24.

———. 1970. The "Keuper" series of the Brisith Trias in the northern Irish Sea and neighbouring areas. *Nature* 226:254–56.

———. 1974. Les évaporites du Trias britannique. *Bull. Soc. Géol. France* 16:708–23.

Watts, A. B. 1981. The U.S. Atlantic continental margin: subsidence history, crustal structure, and thermal evolution. In *Geology of passive continental margins: history, structure, and sedimentologic record,* ed. A. W. Bally et al. p. 2-i-2-75. Am. Ass'n Petrol. Geol. Short Course Education Notes Series 19.

———. 1982. Tectonic subsidence, flexure, and global changes of sea level. *Nature* 297:469–74.

Watts, A. B., and Ryan, W. B. F. 1976. Flexure of the lithosphere and continental margin basins. *Tectonophysics* 36:25–44.

Waugh, B. 1973. The distribution and formation of Permian-Triassic red beds. In *The Permian and Triassic systems and their mutual boundary,* ed. A. Logan and L. V. Hills, p. 678–93. Canadian Soc. Petrol. Geol. Memoir 2.

Weddle, T. K., and Hubert, J. F. 1983. Petrology of Upper Triassic sandstones of the Newark Supergroup in the northern Newark, Pomeraug, Hartford, and Deerfield basins. *Northeastern Geol.* 5:8–22.

Wessel, J. M. 1969. *Sedimentary history of Upper Triassic alluvial fan complexes in north-central Massachusetts.* Dept. of Geol., Univ. of Mass., Contribution 2.

Wessel, J. M.; Hand, B. M.; and Hayes, M. O. 1967. Sedimentary features of the Triassic rocks in northern Massachusetts. In *Guidebook for field trips in the Connecticut Valley of Massachusetts: 59th New England Intercollegiate Geological Conference,* ed. P. Robinson, p. 154–65.

Wheeler, W. H., and Textoris, D. A. 1978. Triassic limestone and chert of playa origin in North Carolina. *J. Sed. Petrol.* 48:765–76.

Whittaker, A. 1980. Triassic salt deposits in southern England. In *Fifth symposium on salt: Northern Ohio Geological Society,* ed. A. H. Coogan and L. Hauber, p. 175–79.

Wilkes, G. P. 1982. Geology and mineral resources of the Farmville Triassic basin, Virginia. *Virginia Minerals* 28:25–32.

Willard, M. E. 1951. *Bedrock geology of the Mount Toby quadrangle, Massachusetts.* U.S. Geol. Survey Map GQ8.

———. 1952. *Bedrock geology of the Greenfield quadrangle, Massachusetts.* U.S. Geol. Survey Map GQ10.

Young, R. S., and Edmundson, R. S. 1954. Oolitic limestone in the Triassic of Virginia. *J. Sed. Petrol.* 24:275–79.

Ziegler, P. A. 1981. Evolution of sedimentary basins in north-west Europe. In *Proceedings of the 2nd Conference on the petroleum geology of the continental shelf of north-west Europe,* ed. L. V. Illing and G. D. Hobson, p. 3–39. London: Institute of Petroleum.

———. 1985. Evolution of the Arctic–North Atlantic rift system [abstract]. *Bull. Am. Ass'n Petrol. Geol.* 69:2047.

Index

Page numbers in *italic* indicate illustrations.

Adhesion ripples, *213,* 219, 225
Alluvial fan: basalts wedge out against, 63; processes and deposits, *118;* sub-environments, 116
Alluvial-fan deposits, 115–39; anomalous deposits, 130–32; clast contours and paleogeography, *138;* response to changes in climate, 133–34, *134;* response to changes in tectonics, 133–36, *135, 136;* as sedimentation controls, 116; study methods, 137–39; tectonic control of 115–16; Triassic-modern comparison, 276
American Journal of Science: establishment by Silliman, 18
Antidunes: recogition of, 83
Appalachia: Barrell's use of, 57
Archaeopteryx, 25
Architectural-element analysis, 271
Armored mud balls, 92–94
Assemblagcs: uses of, 270

Bain, G. W.: interpretations of conglomerates and shales, 66
Barrell, J., 55–59; concepts of geological time, 59; concepts of nonmarine sedimentation, 56–57; reconstruction of climate, 58; redbed origins, 57–58; synsedimentary faulting, 58–59; use of Appalachia, 57; use of facies, 56–57
Basalt: Hartford basin, 45–46, 80–82; theories of origin, 15, 18
Basalt flows: Aquitaine basin, 243; effects on lacustrine deposits, 205-7,*206;* effects on marine-influenced deposits, 228–29, *228*
Bay of Fundy: analogue to Hartford basin, 29, 40; Emerson's use of, 49–52; Lyell's comparison with, 29; map, *40;* Russell's use of, 39–41
Border Conglomerate: significance of, in Hartford basin, 58–59, 63
Braided-river deposits, 141–48; characteristics, *143;* outcrop sketch, *142;* sheetlike sandstones, *144;* vertical trends in, 145–46; very large river models, 146–47
Broad-terrane hypothesis: disputed by Klein, 85; Krynine's arguments against, *71;* map, *42;* origin by Russell, 41–43; Sanders's use of, 80; use of, by Wheeler, 66–67
Bromine: evidence for depositional environment, 233, 246

Caliche: in overbank deposits, 153–55, 158, 159
Carboniferous: correlations with the Hartford group, 22
Catastrophism: H. D. Rogers, 30, 32
Chapman, R. W.: study of basalt in the Hartford basin, 80–81
Cherry Valley basin: discovery of, 78; map, 8
Climate: Barrell's reconstruction of, 58; controlled by paleogeography, 263–64, 264; controls sedimentation, 261–65; Krynine's reconstruction of, 71–74; Russell's warm-humid theory, 43–44; Triassic, 114
Coal formation: in the Hartford basin, 22
Cooper, T.: on origin of basalts, 18
Crosby, W. O.: origin of redbeds, 58
Cross sections: uses of, 268
Crumb fabrics: in marine-influenced deposits, 237; in playa deposits, 215
Cyclic deposition: marine-influenced deposits of England, 236–37, 240
Cyclic lacustrine deposition, 89, 200–204

Dana, J. D., 37–39; concepts of sedimentary structures, 37; glacial deposition, 38; seaward repositioning of coastline, 38; theory of geosynclines, 38
Davis, W. M., 45–49; dispute with Dana, 45; map of Hartford basin, 47; nonmarine interpretations of, 48; stratigraphic nomenclature of, 47; structural interpretations of, 46, 48–49, 49
Deane, J.: conflict with Hitchcock, 25
De Boer, J.: paleomagnetic studies by, 81–82
Debris-flow deposits, 117, 118–19, 120, 121; characteristics, 122; relationship of clast size to bed thickness, 119
Deep River basin: stratigraphic position of coals, 169
Deerfield basin: location of, 8; 1818 geological map of, 20; 1844 geological

map of, 24; 1898 geological map of, 51; 1951 geological map of, 76
Desiccation cracks, 156; Hitchcock's recognition of, 28; in playa deposits, 217, 218; Silliman's description of, 19
Dikes: orientation of, 113, 113
Dinornis, 25
Dinosauria: creation of class of, 25

East Berlin Formation: general description, 12
Emerson, B. K., 49–52, 60–62; depositional model used by, 50–52, 60–62; disregard for structure in Hartford basin, 50; use of Bay of Fundy analogue, 49–52; 1898 map of Deerfield and Hartford basins, 51
Eolian deposits, 248–59; adhesion ripples, 255; age and location, 249; associated facies, 257; barchan dunes, 252, 253; characteristics, 251; climbing-ripple translatent strata, 251; controls on facies, 257; correlation of mica and rounded grains, 254–55; deflation surfaces, 251, 252, 255; dome-shaped dunes, 253–54, 254; grain-fall and grain-flow laminae, 250, 251; interdune deposits, 255–56; study methods, 258; transverse dunes, 254; Triassic–modern comparison, 281
Era of state surveys, 20
Estuarine model: early concepts of, in the Hartford basin, 26; Dana's support of, 38; Emerson's use of, 60–62; Lyell's inferences, 29; Russell's use of, 39–41
Evaporites: concentric patterns in Morroccan basins, 223–25, 225, 230; correlation of halite beds, 235; in England, 231–41; in lacustrine deposits, 199–200; polygons in halite, 238, 239

Facies: Barrell's use of, 56–57; early usage of concept, 45; lack of use by

Roberts, 64, *64;* Russell's use of, 44–45

Facies models: uses of, 272; Triassic–modern comparison, 275–81

Fence diagrams: uses of, 268

Filtering and interpretation of data, 108–9

Fish fossils: Redfields' study of, 35–36

Flexural bulges: controls on rift-basin sedimentation, *262;* possible effects on Hartford basin, 102

Fluvial deposits, 140–64; channels in mudstones, *161;* climatic and tectonic controls on, 157–58; on fans, 117; study methods, 162–64; Triassic–modern comparison, 276–78

Fluvial models: use of, in Hartford basin, 30, 91–94

Fossils: uses of, 273

Foye, W. G.: recognition of pillow basalts, 65

Geochemistry: uses of, 273; lake waters, 198

Geological time: Barell's expansion of, 59

Geosyncline: Dana's theory of, 38; Kay's ideas of, 39; Krynine's concept of its control on petrology, 74

Glacial deposition: Dana's model of, 38

Grain-fall deposits, *250,* 251

Grain-flow deposits, *250,* 251

Gravity-loading structures, *155*

Gravity study: Hartford basin, 82

Halite: correlation of bedding, *235;* deposition in Morocco, 224–25, *224;* polygons and vertical cracking, 236, *238,* 239; pseudomorphs in Hartford basin, *61,* 96

Hampden basalt: general description, 12

Hartford basin: age of strata, 35, 55; climate, 88, 94, 100; climate and cycles, 89; coal, 17; cross sections of, *10, 57,*

87; early concepts of structure, 48–49, 63; inherited structure, 86; interaction of tectonics, sedimentation, and paleoslope, 101; map, 8–9; paleogeography, *93;* paleomagnetic studies, 81–82; stratigraphy, *7,* 7–10, 35, *69, 72;* structure, 10, 80, *82;* suggestions for research in, 100–103; synsedimentary faulting, 58–59; tectonics, 12–14; unconformity in, *87,* 87–88, 101–2; 1809 geological map, *17;* 1823 geological map, *21;* 1842 geological map, *34;* 1844 geological map, *24;* 1898 geological maps, *47, 51;* 1950 geological map, *70;* 1985 geological map, *79*

Hartford and Deerfield basins: separation of, 11

Hitchcock, E., 19–29; concepts of sedimentary structures, 27–28; conflict with Deane, 25; contributions to ichnology, 23–26; geological support for religion, 27; 1823 cross section of the Hartford basin, *23;* 1823 geological map, *21*

Hitchcock Volcanics, 22

Hobbs, W. H.: study of the Pomperaug outlier, 52–53

Holyoke basalt: general description, 12

Hubert, J. F.: recognition of lacustrine sedimentation by, 88–91; use of fluvial models in Hartford basin by, 91–94

Huttonian theory, 15

Ichnology: Hitchcock's contributions to, 23–26

Inclined deposition: theory of, 26

Iron: mining of, 3

Isotope studies: of playa evaporities, 219–20

Kay, M.: theory of geosynclines, 39n

Kelvin, Lord: age of earth, xvi

Klein, G. de V.: isolated basins, 85; lacustrine interpretations of, 85, 211

Krynine, P. D., 68–74; climatic significance of redbeds, 71–74; paleogeographic interpretations of, *72;*
petrology as a function of geosynclinal cycle, 74; redbed deposition,
73–74; stratigraphic divisions of
Hartford group by, *72;* theory of
Pomperaug outlier, 69–71, *71;* use of
petrology by, 69–71; 1950 map of
Hartford basin, *70*

Lacustrine carbonates, 187–90; oncolites,
190, *190;* oolites, 188; stromatolites,
188, *206;* teepee structures, 188, *189*
Lacustrine conditions: origin of, 100–101,
181–82
Lacustrine cycles: Hartford basin, *90*
Lacustrine deep-water deposits, 193–98;
lateral extent, *195,* 197; seasonal
varves, 194; thicken toward border
fault, *194;* turbidites, 197, *198;* varved
mudstones, *196*
Lacustrine deposits, 180–209; associated
facies, 204; authigenic minerals, 199;
black mudstones as shallow-water
facies, 195; characteristics, *181;* chemical vs. detrital cycles, 200, *201;* cyclic
deposition, 200–204; effects of basalt
flows on, 205–7, *206;* geochemistry of
water, 198–200; invertebrate fossils,
199; maximum water depths, 193–97;
modified cyclic deposition, 202–4,
203; presence of evaporites, *61,* 96,
199–200; rippled bedding, *192;* small
lakes, 182; study methods, 207–9;
tempestites, *184,* 185; Triassic–modern
comparison, 278; turbidities, *191,* 197,
198
Lacustrine geochemistry: controlled by
climate, 263
Lacustrine shallow-water clastics, 190–93
Lacustrine shallow-water deposits, 187–93
Lacustrine shoreline deposits, 183–87;
beaches, 184–86, *185, 186;* deltas,

183–84; mudflats, 186–87; wave-cut
terraces, 186
Lakes: origin of, in Hartford basin, 89
Land bridge: hypothesis of Lull, 60;
hypothesis of Schuchert, 60
Lateral fractionation: evaporites, 224,
225, 244
Longwell, C. R., 62–65
Lull, R. S., 59–60; proposed land
bridge, 60
Lyell, C.: arguments against primary
depositional dip, 31; influence on
Hitchcock, 27; support for estuarine
model, 29; uniformitarianism, 32

Maclure, W., 16; on speculative geologic
models, 108; 1809 map, *17*
Mapping: uses of, 267
Marine deposition: early concepts of, in
the Hartford basin, 26
Marine-influenced deposits, 221–47;
Aquitaine basin, 242–43, *243;*
associated facies, 244–45; bromine
in evaporites, 233, 246; cyclic deposition, 236–37, *240;* deep-water halite,
224–25, 234–36, 242; effects of basalt
flows, 228–29, *228;* in England, 231–42;
in Greenland, 244; lateral fractionation
of evaporites, 244; in Morocco,
223–31; nonmarine interpretations,
234; paralic vs. shallow-marine, 224–31,
225; polygons and vertical cracks in
halite, 236, *238, 239;* sabkha deposits,
231–32; salina model, 225, 236–41;
source of brines, 232–34; study methods, 245–47; tidal deposits, 233, 241–42,
244; transgressive facies, 241–42;
Triassic–modern comparison, 280–81
Meandering-river deposits, 148–50; lateral accretion surfaces, 148–49, *149;*
sandstone-mudstone ratio, 149, *150*
Middleton basin: discovery of, 96
Morocco: paleodrainage, *131;* trans-
Atlantic correlations with, 86

Mud curls: in playa deposits, 215–16, *217*

Mud flows: on alluvial fans, 119

Mudstone facies: study of, in Hartford basin, 102–3

Nodular dolomite/anhydrite, *189*

Nonmarine environments: first interpretations of, in Hartford basin, 48

Nonmarine sedimentation: Barrell's use of, 56–57

Newark Group: Russell's use of, 45

Newark Supergroup: named, 6

New Haven Arkose: general description, 11

New Red Sandstone: in the Hartford basin, 22

Northfield basin: map, *8,* 11

Old Red Sandstone, 16

Oncolites, 190, *190*

Oolites, 188

Ornithichnites, 25

Overbank deposits, 150–57; secondary modification of, 152–57, *154, 155*

Paleomagnetic studies: Hartford basin, 81–82

Paleontology (vertebrate): as interpreted by Lull, 59–60

Paleosols: in overbank deposits, 153–55, *158, 159, 160;* nodules reworked as fluvial lags, *160;* recognition of, in Hartford basin, 94

Palisades disturbance, 55

Paludal deposits, 165–79; age, 165; coal isopach maps, *171;* correlation of coals, *172;* cyclic deposition, *177;* in Deep River basin, 168–73; depositional environments, 173, 174, *174;* environments of deposition, 177; as evidence for climate, 167–68; faulted "facies changes," 174, *175;* location, *166;* location in stratigraphic sequence,

168, *169,* 173; study methods, 178–79; tectonic controls on, 176–77; as tectonic indicator, 168; trace minerals, 173; Triassic–modern comparison, 278

Palynology: dating Hartford group with, 86–87

Percival, J. G., 33–35; as a collector of data, 33; stratigraphic nomenclature of, 35; unpublished interpretive ability of, 35; 1842 geological map by, *34*

Petrology: Krynine's use of, 69–71; uses of, 274–75

Pillow basalts, *65;* in Morocco, 228, *228;* recognition of, in Hartford basin, 65–66

Plate tectonics: application to rift basins, 85–86; controls sedimentation, 260–61, *262*

Playa deposits, 210–20; associated facies, 216; desiccation cracks, *218;* development of paleosols on, 214–15, *214, 215;* evaporites in, 219–20; inorganic precipitation, 212; modifications to, 213–16; mudflat deposits, 211–13, *212;* study methods, 219–20; trace fossils in, 213; Triassic–modern comparison, 279; variable environments, 210–11

Playa model: use in Hartford basin, 91

Pomperaug outlier, 52–53, 63, 66–67, 92; maps, *9, 17;* study of, by Hobbs, 52–53; study of, by Krynine, 69–71, *71*

Potash minerals: Morocco, 224, 229

Primary depositional dip: application of theory to Hartford basin, 29–32; arguments against, 31; evidence for, 30–31

Progressive development: theory of, 26

Raindrop impressions: Lyell's recognition of, 29

Redbeds: Barell's theory of, 57–58; controversy over, 67–68; Crosby's theory of, 58; Krynine's concepts of, 71–74; origin and significance of, 83; Russell's theories, 43; transported red sediment, 43–44

Redfield, J. H. and Redfield, W. C.: Triassic-Jurassic age of fish, 35–36

Religion: supported by geology, 27

Rift basins: location map, *112;* orientation of, 111–13, *113;* regional setting, 111–14

Rifting: processes of, 113–14

Rift valleys: first analogies to Hartford basin, 80

Ripple index, 256

Roberts, J. K.: misinterpretation of Hartford basin, 64, *64*

Rogers, H. D., 29–32; and catastrophism, 30; primary depositional dip, 30–32

Russell, I. C., 39–45; broad-terrane hypothesis of, 41–43; support for estuarine theory, 39–43; theories of climate, 43–44; theories of red color, 43; use of facies concepts, 44–45; use of term *Newark Group,* 45

Russell, W. L. 63

Sabkha deposits: Morocco, 231, *232*

Sanders, J. E., 83–85; recognition of lacustrine deposits by, 84–85; structural studies by, 80

Scientific method: discussion of, 281–85

Scoyenia: in lacustrine deposits, 191; in overbank deposits, 152, 155

Sea level: controls sedimentation, *265,* 265–66

Secondary Formation, 16

Sedimentological systems: scales of, 109

Sedimentology: division of the history of, 99; origin of term, 55; quantification of, 68

Shuttle Meadow Formation: general description, 12

Sieve deposits, 116, 119–20, *123*

Silliman, B.: appointment to Yale, 3, 15; constrained by religion, 15; geological

support for religion, 27; on origins of basalt, 15–16

Silliman, B. Jr.: application of theory of primary depositional dip to Hartford basin, 30

Sheetflood deposits, *126;* characteristics, *123;* general description, 117; graded bedding in, *125;* horizontal-parallel bedding in, 124; imbrication of clasts, *124;* midfan, 120–25; recognition of, in Hartford basin, 94–95; terminology, 116

Sheetflood (fan-toe) deposits, 125–30; general description, 117; graded bedding in, *127, 128;* horizontal-planar bedding, *129, 130*

Sorby, H. C., 37

Stratigraphic nomenclature: comparison of different authors' terms in Hartford basin, *69;* correlation between Hartford and Deerfield basins; 78; Krynine's terminology for Hartford basin, 69

Stromatolites, 188, *206*

Study methods: assemblages, 270–72, *271;* categorization and relationships, 282–83; consilient theories, 284; cross sections, 268; facies models, 272; fence diagrams, 268, *269;* fossils and trace fossils, 273; geochemistry, 273; mapping, 267–68; petrography and grain-size analysis, 274–75; present as key to past, 282; significance of anomalies, 283–84; vertical profiles, 268

Subaqueous sedimentation: origins of concepts, 18–19

Talcott basalt, 12

Tectonics: controls on local sedimentation, 266–67

Teepee structures, 188, *189*

Theories: uses of, 98–99

Tidal deposits: England, 241–42

Tidal deposition in Hartford basin: Bay

of Fundy analogue, 41; Emerson's
concept of, 50–52, 62; Hitchcock's
concept of, 26; Russell's concept of,
39–40
Triassic: climate, 114; origin of term, 17
Triassic–Jurassic boundary: placement
in Hartford basin, 87
Trap rocks, 16
Turbidites, *191,* 197, *198*

Van Houten, F. B.: cyclic lacustrine
sedimentation, 89, 200–201, *201*

Varves: recognition of, in Hartford basin,
89
Vertical profiles: uses of, 268

Wernerian theory, 15
Wessel, J. M.: study of conglomerates by,
82–83
Wheeler, G.: structural interpretation of,
66–67
Willard, M. E.: map of Deerfield basin,
76; revision of Emerson's map by,
77